Samuel Butler and the Science of the Mind
Evolution, Heredity and Unconscious Memory

LEGENDA

LEGENDA is the Modern Humanities Research Association's book imprint for new research in the Humanities. Founded in 1995 by Malcolm Bowie and others within the University of Oxford, Legenda has always been a collaborative publishing enterprise, directly governed by scholars. The Modern Humanities Research Association (MHRA) joined this collaboration in 1998, became half-owner in 2004, in partnership with Maney Publishing and then Routledge, and has since 2016 been sole owner. Titles range from medieval texts to contemporary cinema and form a widely comparative view of the modern humanities, including works on Arabic, Catalan, English, French, German, Greek, Italian, Portuguese, Russian, Spanish, and Yiddish literature. Editorial boards and committees of more than 60 leading academic specialists work in collaboration with bodies such as the Society for French Studies, the British Comparative Literature Association and the Association of Hispanists of Great Britain & Ireland.

The MHRA encourages and promotes advanced study and research in the field of the modern humanities, especially modern European languages and literature, including English, and also cinema. It aims to break down the barriers between scholars working in different disciplines and to maintain the unity of humanistic scholarship. The Association fulfils this purpose through the publication of journals, bibliographies, monographs, critical editions, and the MHRA Style Guide, and by making grants in support of research. Membership is open to all who work in the Humanities, whether independent or in a University post, and the participation of younger colleagues entering the field is especially welcomed.

ALSO PUBLISHED BY THE ASSOCIATION

Critical Texts
Tudor and Stuart Translations • *New Translations* • *European Translations*
MHRA Library of Medieval Welsh Literature

MHRA Bibliographies
Publications of the Modern Humanities Research Association

The Annual Bibliography of English Language & Literature
Austrian Studies
Modern Language Review
Portuguese Studies
The Slavonic and East European Review
Working Papers in the Humanities
The Yearbook of English Studies

www.mhra.org.uk
www.legendabooks.com

STUDIES IN COMPARATIVE LITERATURE

Editorial Committee
Chairs: Dr Emily Finer (University of St Andrews)
and Professor Wen-chin Ouyang (SOAS, London)

Dr Ross Forman (University of Warwick)
Professor Angus Nicholls (Queen Mary, University of London)
Dr Henriette Partzsch (University of Glasgow)
Dr Ranka Primorac (University of Southampton)

Studies in Comparative Literature are produced in close collaboration with the British Comparative Literature Association, and range widely across comparative and theoretical topics in literary and translation studies, accommodating research at the interface between different artistic media and between the humanities and the sciences.

ALSO PUBLISHED IN THIS SERIES

20. *Aestheticism and the Philosophy of Death: Walter Pater and Post-Hegelianism*, by Giles Whiteley
21. *Blake, Lavater and Physiognomy*, by Sibylle Erle
22. *Rethinking the Concept of the Grotesque: Crashaw, Baudelaire, Magritte*, by Shun-Liang Chao
23. *The Art of Comparison: How Novels and Critics Compare*, by Catherine Brown
24. *Borges and Joyce: An Infinite Conversation*, by Patricia Novillo-Corvalán
25. *Prometheus in the Nineteenth Century: From Myth to Symbol*, by Caroline Corbeau-Parsons
26. *Architecture, Travellers and Writers: Constructing Histories of Perception*, by Anne Hultzsch
27. *Comparative Literature in Britain: National Identities, Transnational Dynamics 1800-2000*, by Joep Leerssen
28. *The Realist Author and Sympathetic Imagination*, by Sotirios Paraschas
29. *Iris Murdoch and Elias Canetti: Intellectual Allies*, by Elaine Morley
30. *Likenesses: Translation, Illustration, Interpretation*, by Matthew Reynolds
31. *Exile and Nomadism in French and Hispanic Women's Writing*, by Kate Averis
32. *Samuel Butler against the Professionals: Rethinking Lamarckism 1860–1900*, by David Gillott
33. *Byron, Shelley, and Goethe's Faust: An Epic Connection*, by Ben Hewitt
34. *Leopardi and Shelley: Discovery, Translation and Reception*, by Daniela Cerimonia
35. *Oscar Wilde and the Simulacrum: The Truth of Masks*, by Giles Whiteley
36. *The Modern Culture of Reginald Farrer: Landscape, Literature and Buddhism*, by Michael Charlesworth
37. *Translating Myth*, edited by Ben Pestell, Pietra Palazzolo and Leon Burnett
38. *Encounters with Albion: Britain and the British in Texts by Jewish Refugees from Nazism*, by Anthony Grenville
39. *The Rhetoric of Exile: Duress and the Imagining of Force*, by Vladimir Zorić
40. *From Puppet to Cyborg: Pinocchio's Posthuman Journey*, by Georgia Panteli
41. *Utopian Identities: A Cognitive Approach to Literary Competitions*, by Clementina Osti
43. *Sublime Conclusions: Last Man Narratives from Apocalypse to Death of God*, by Robert K. Weninger
44. *Arthur Symons: Poet, Critic, Vagabond*, edited by Elisa Bizzotto and Stefano Evangelista
45. *Scenographies of Perception: Sensuousness in Hegel, Novalis, Rilke, and Proust*, by Christian Jany
46. *Reflections in the Library: Selected Literary Essays 1926–1944*, by Antal Szerb
47. *Depicting the Divine: Mikhail Bulgakov and Thomas Mann*, by Olga G. Voronina
48. *Samuel Butler and the Science of the Mind: Evolution, Heredity and Unconscious Memory*, by Cristiano Turbil
49. *Death Sentences: Literature and State Killing*, edited by Birte Christ and Ève Morisi
50. *Words Like Fire: Prophecy and Apocalypse in Apollinaire, Marinetti and Pound*, by James P. Leveque

Samuel Butler and the Science of the Mind

Evolution, Heredity and Unconscious Memory

Cristiano Turbil

LEGENDA
Studies in Comparative Literature 48
Modern Humanities Research Association
2020

Published by Legenda
an imprint of the Modern Humanities Research Association
Salisbury House, Station Road, Cambridge CB1 2LA

ISBN 978-1-78188-553-6 (HB)
ISBN 978-1-78188-554-3 (PB)

First published 2020

All rights reserved. No part of this publication may be reproduced or disseminated or transmitted in any form or by any means, electronic, mechanical, photocopying, recording or otherwise, or stored in any retrieval system, or otherwise used in any manner whatsoever without written permission of the copyright owner, except in accordance with the provisions of the Copyright, Designs and Patents Act 1988, or under the terms of a licence permitting restricted copying issued in the UK by the Copyright Licensing Agency Ltd, Saffron House, 6–10 Kirby Street, London EC1N 8TS, England, or in the USA by the Copyright Clearance Center, 222 Rosewood Drive, Danvers MA 01923. Application for the written permission of the copyright owner to reproduce any part of this publication must be made by email to legenda@mhra.org.uk.

Disclaimer: Statements of fact and opinion contained in this book are those of the author and not of the editors or the Modern Humanities Research Association. The publisher makes no representation, express or implied, in respect of the accuracy of the material in this book and cannot accept any legal responsibility or liability for any errors or omissions that may be made.

Trademark notice: Product or corporate names may be trademarks or registered trademarks, and are used only for identification and explanation without intent to infringe.

© Modern Humanities Research Association 2020

Copy-Editor: Dr Marie Isabel Matthews-Schlinzig

CONTENTS

	Acknowledgements	ix
	Chronology of Samuel Butler	x
	Introduction	1
1	Butler and the Science of the Mind: A Pan-European Discussion	13
2	Evolution: From Literature to Science and Back Again	35
3	The Rise and Fall of Butler's Fame	61
4	An Amateur among the Professionals	85
	Conclusion: Another Story to Tell	121
	Bibliography	131
	Index	141

In memory of my Father
(1954-2019)

ACKNOWLEDGEMENTS

Writing this book on Samuel Butler's scientific ideas on evolution and the mind has been a long process which started many years ago when I moved to England for my PhD. During this time, I have benefited greatly from the help and advice of archivists and librarians at various national and international institutions. These include the British Library in London; St. Johns College Library at the University of Cambridge; the Cambridge University Library, Department of Manuscripts and University Archives; the Bodleian Library at Oxford University; the Wellcome Library in London; the Chapin Library of Rare Books at Williams College in Cambridge, USA and the Biblioteca Fardelliana in Trapani, Italy.

I would like to thank my supervisors, peers and friends during my time as a PhD student at the University of Kent. I also wish to thank all of my colleagues at King's College London and University College London for their interest and input throughout the years. In particular, a special thanks to Prof. Elinor Shaffer who strongly believed in this project from the start and always supported my work on Butler's writing on science and its pan-European dimension.

I am also grateful to the editorial committees of two journals. Parts of Chapters 1 and 4 draw on my earlier articles 'Making Heredity Matter: Samuel Butler's Idea of Unconscious Memory', *Journal of the History of Biology*, 51 (2018), 7–29, and 'In between mental evolution and unconscious memory: Lamarckism, Darwinism and professionalism in late Victorian Britain', *Journal of the History of the Behavioral Sciences*, 53 (2017), 347–63.

Finally, I thank my Italian and British families, and Rachel in particular for continuously supporting me.

c.t., Uckfield, July 2020

CHRONOLOGY OF SAMUEL BUTLER

❖

This chronology of Butler's life and works is based on the 'Biographical Statement' which can be found in Butler, *The Notebooks*, pp. 1–8.

1835	On 4 December, Samuel Butler is born at Langar Rectory, Nottingham, UK.
1843–44	First travels in Italy with his family, visiting Rome and Naples.
1846–48	Starts school at Allesley, near Coventry.
1848–54	Attends Shrewsbury School.
	Goes to Italy for the second time with his family.
	First encounter with the music of George Frideric Handel.
1854	Enrolls at St John's College, Cambridge, UK.
1858	Graduates from Classical Tripos at St John's College, Cambridge.
	Moves to London to begin preparation for ordination.
1859	Decides to abandon the religious life and moves to New Zealand. There, his aim is to become a sheep farmer in Canterbury Province.
	Charles Darwin publishes *On the Origin of Species by Means of Natural Selection, or the Preservation of Favoured Races in the Struggle for Life*.
1862	On 20 December, Butler publishes an unsigned dialogue in the New Zealand periodical *The Press* entitled 'Darwin on the Origin of Species: A Dialogue'.
	Charles Darwin publishes *On the Various Contrivances by which British and Foreign Orchids are Fertilised by Insects*.
1863	Butler publishes *A First Year in Canterbury Settlement*.
1863	Butler writes and publishes in *The Press*: 'Darwin among the Machines', a letter signed 'Cellarius'.
1864	Decides to sell his farm and return to England in the company of a friend he made in New Zealand: Charles Paine Pauli. In London, he settles at 15 Clifford's Inn, where he decides to start a new career and to become a painter. Henry Festing Jones reports that Butler enrols at several arts schools including the South Kensington Art School, Cary's Art School, and the Heatherley School of Fine Art.
	Family Prayers: a painting by Butler.
1865	Butler publishes a second short article about evolution in *The Press*: 'Lucubratio Ebria'.
	Butler publishes *The Evidence for the Resurrection of Jesus Christ as Contained in the Four Evangelists Critically Examined* — a pamphlet written in New Zealand.
1868	Charles Darwin publishes *The Variation of Animals and Plants under Domestication*.
1869–70	Travels to Italy, especially the north of the country (in particular Piedmont and Lombardy).
1870	First meeting with Eliza Mary Ann Savage.

	Alfred R. Wallace publishes *Contributions to the Theory of Natural Selection: A Series of Essays*.
	Ewald Hering delivers at the University of Prague the lecture 'Über das Gedächtnis als eine allgemeine Funktion der organisierten Materie' ('On Memory as a Universal Function of Organized Matter').
1871	Charles Darwin publishes *The Descent of Man, and Selection in Relation to Sex*.
1872	Butler publishes his first novel: *Erewhon, or, Over the Range* — a satire.
	Charles Darwin publishes *The Expression of Emotions in Man and Animals*.
	St. George Mivart publishes *On the Genesis of Species*.
1873	*Erewhon* is translated into Dutch.
	Butler publishes *The Fair Haven*.
	Mivart publishes *Man and Apes: An Exposition of Structural Resemblances and Differences Bearing upon Questions of Affinity and Origin*.
	Théodule Ribot publishes *L'Hérédité: Étude psychologique*.
1874	Paints *Mr. Heatherley's Holiday*, which, according to Henry Festing Jones, is Butler's most important oil painting.
	Mivart publishes *An Examination of Mr. Herbert Spencer's Psychology*.
	Butler travels to Montreal, Canada.
1875	Charles Darwin publishes *Movement and Habits of Climbing Plants*.
1876	Meets Henry Festing Jones.
1877	Butler publishes his first book on science: *Life and Habit: An Essay after a Completer View of Evolution*, dedicated to Charles Paine Pauli. Although dated 1878, the book is published on Butler's birthday, 4 December 1877.
1878	Butler publishes the poem 'A Psalm of Montreal' in the *Spectator*.
1879	Butler publishes his second book on evolution: *Evolution, Old and New: or, The Theories of Buffon, Dr. Erasmus Darwin and Lamarck, as Compared with that of Charles Darwin* Butler publishes *A Clergyman's Doubts* and the articles (eight in total) 'God the Known and God the Unknown' in the *Examiner*.
	Erewhon is translated into German.
	Charles Darwin publishes 'Preface and "A Preliminary Notice"' in *Erasmus Darwin* by Ernst Krause.
1880	Butler publishes his third book on evolution: *Unconscious Memory: A Comparison Between the Theory of Dr. Ewald Hering, Professor of Physiology in the University of Prague, and the 'Philosophy of the Unconscious' of Dr. Edward von Hartmann, with Translations from both these Authors, and Preliminary Chapters Bearing upon 'Life and Habit', 'Evolution, Old and New', and Mr. Charles Darwin's Edition of Dr. Krause's 'Erasmus Darwin'*.
1881	Butler publishes his first book on Italian art: *Alps and Sanctuaries of Piedmont and the Canton Ticino*.
	Théodule Ribot publishes *Les Maladies de la mémoire*.
1882	A new edition of *Evolution, Old and New* is published with a short preface alluding to the recent death of Charles Darwin, an appendix, and an index.
1883	Starts to compose music in the style of Handel.
	George Romanes publishes *Mental Evolution in Animals: With a Posthumous Essay on Instinct by Charles Darwin*.

1884	Butler publishes *Selections from Previous Works* with 'A Psalm of Montreal' which includes the long pamphlet 'Remarks on G. J. Romanes' *Mental Evolution in Animals*'.
1885	Death of Eliza Mary Ann Savage.
	Together with Henry Festing Jones, Butler publishes his first music composition: *Gavottes, Minuets, Fugues: And Other Short Pieces for the Piano*.
1886	George Romanes publishes *Physiological Selection: An Additional Suggestion on the Origin of Species*.
1887	Butler publishes his final and most controversial book on science: *Luck, or Cunning as the Main Means of Organic Modification? An Attempt to Throw Additional Light upon Charles Darwin's Theory of Natural Selection*.
	Publication of *Autobiography* of Charles Darwin (ed. by his son Francis Darwin).
	Publication of *Life and Letters of Charles Darwin* (ed. by Francis Darwin).
1888	Butler publishes his second book on Italian art: *Ex Voto: An Account of the Sacro Monte or New Jerusalem at Varallo-Sesia*.
	With Henry Festing Jones, Butler composes 'Narcissus: A Cantata in the Handelian Form', which remains unpublished.
	In this and the two following years, Butler contributes some articles to the *Universal Review*, most of which were republished after his death as *Essays on Life, Art, and Science* (1904).
1889	Alfred R. Wallace publishes *Darwinism: An Exposition of the Theory of Natural Selection with Some of its Applications*.
1892	Butler publishes *The Humour of Homer*, a lecture delivered at the Working Men's College, Great Ormond Street, London, 30 January 1892, reprinted with a preface and additional matter from the *Eagle*.
	Visits Sicily for the first time to collect evidence in support of his theory identifying the Scheria and Ithaca of the *Odyssey* with Trapani and the neighbouring Mount Erice.
1893	Butler publishes 'L'origine Siciliana dell'Odissea' in *Rassegna della Letteratura Siciliana*. Later translated into English as 'On the Trapanese Origin of the Odyssey'.
1894	Butler's *Ex Voto* is translated into Italian by Angelo Rizzetti.
	Butler publishes 'Ancora sull'origine dell'Odissea', extracted from the *Rassegna della Letteratura Siciliana*.
1895	Goes to Greece and the Troad to make up his mind about the topography of the *Iliad*.
1896	Butler publishes *The Life and Letters of Dr. Samuel Butler in so far as They Illustrate the Scholastic, Religious and Social Life of England from 1790–1840*.
1897	Butler publishes *The Authoress of the Odyssey*.
1897	Death of Charles Paine Pauli.
1898	Butler translates Homer's *Iliad* into English.
1899	Butler publishes *Shakespeare's Sonnets Reconsidered and in Part Rearranged: With Introductory Chapters, Notes and a Reprint of the Original 1609 Edition*.
1900	Butler translates Homer's *The Odyssey* into English.
1901	Butler publishes his second novel: *Erewhon Revisited*.
1902	On 18 June, Butler dies in London.

INTRODUCTION

> Man is a walking tool-box, manufactory, workshop and bazaar worked from behind the scenes by someone or something that we never see. We are so used to never seeing more than the tools, and these work so smoothly, that we call them the workman himself, making much the same mistake as though we should call the saw the carpenter. The only workman of whom we know anything at all is the one that runs ourselves and even this is not perceivable by any of our gross palpable senses.
>
> The senses seem to be the link between mind and matter — never forgetting that we can never have either mind or matter pure and without alloy of the other.
>
> SAMUEL BUTLER, *The Notebooks*[1]

In recent years, scholarly interest in the work of the Victorian Samuel Butler has grown steadily: his controversial personality and ideas on art, literature, and science have become part of an ongoing discussion that has equally fascinated historians and literary scholars. However, his contribution to the evolutionary debate has been generally overlooked. Indeed, since the early twentieth century, scholars have neglected the significance of Butler's writings on science and his idea of 'unconscious evolution'. Despite some recent works, for example James Paradis' *Samuel Butler, Victorian against the Grain* (2007) and David Gillott's *Samuel Butler against the Professionals* (2015), little has been published on Butler's writings on evolution and the mind. There are various reasons for this gap in the scholarship.

First, Butler's writings on science have never been considered as a serious attempt to discuss evolution — until recently. However, as will be argued in this volume, Butler's engagement with science does have merit and can help us to uncover an additional layer to the European history of evolution. Second, new research on Lamarckism and its role in the development of the study of the mind in the late nineteenth and early twentieth centuries have created the ideal conditions for a general re-evaluation of Butler's work and theories. In this respect, to fully understand the significance and place of Butler's science of the mind in both the psychological and evolutionary debate of the late nineteenth and early twentieth centuries, his work cannot be discussed in isolation. This is the main aim of this monograph: to reframe Butler's writing on science in relation to the psychological and physiological contexts of late nineteenth-century British and continental debates. In order to do so, a very brief contextualization of some of the existing studies on the topic, but also more generally about what historians have been writing on the significance of evolutionary theory in the period, becomes necessary. This exposition will provide an overview of some of the approaches that scholars have taken in the last few decades.

Over the past twenty years, historians of science and literary scholars, interested in the relationship between culture and natural science have produced a rich scholarship on the history of evolution in the Victorian period. They explored how Charles Darwin's theory influenced the scientific but also more generally the cultural debates of the time.[2] Most of what has been written centres on one major topic, which equally obsessed nineteenth-century biologists, philosophers, and writers: the desire to understand, discuss, and explore the place and significance of evolution in the cultural, social, and scientific debates of their time. The interest in evolution involved the whole society. From the professional to the 'man on the street', everyone was fascinated by nature and its secrets.

This fascination for nature and its history, however, started long before the publication of Darwin's *On the Origin of Species* (1859). Ralph O'Connor's *The Earth on Show* (2007) provides a clear account of how the public engaged with geology and Earth's history since the very early nineteenth century. What O'Connor suggests is the importance of looking at how natural science became an object to be put on display, to be discussed in the public sphere, and to be represented in literary and poetical forms by various writers. In a similar manner, James Secord's *Victorian Sensation* (2000) looks at the reception of Robert Chambers' *Vestiges of the Natural History of Creation* (1844), exploring how the publication of a single book changed the perspective on and interest in science for a relatively large portion of the Victorian public. Both O'Connor and Secord invite the reader to observe the growing fascination that Victorians developed for the natural world and its history. Indeed, as these scholars have observed, the audience and the practice of communication of science became key aspects of the Victorian evolutionary debate. In this respect, the nineteenth-century fascination for the natural world needs to be taken into serious account, and the history of the popularization of science has to be analysed as both product and shaper of Victorian science. Only in this way can we understand why the debate about science and evolution became a phenomenon of such proportions.

In regard to the nineteenth-century fascination for nature, it is important to further stress that among the Victorians no one was exempt. One such example of this Victorian passion for natural history can be found in an anecdote concerning Queen Victoria herself. In 1842, five years before London Zoo fully opened to the general public in 1847, Queen Victoria had the chance to visit it for the first time, and while observing the animals she became impressed by one in particular: the orang-utan (genus *Pongo*). London Zoo had opened for scientific purposes in 1827. In early 1830, animals were transferred to the zoo from the Tower of London. The aim was to create a space where the general public could experience and observe the behaviours of exotic animals. During her visit, the Queen declared: 'the Orang Outang is too wonderful ... he is frightfully, and painfully, and disagreeably human.'[3] Queen Victoria was impressed by the incredible resemblance between the two species. At the time, the possibility of considering the orang-utan as the link between animals and humans was far from being scientifically accepted. In 1838, a young Charles Darwin (1809–1882), at the beginning of his work on evolution, wrote in his notebooks: 'Man in his arrogance thinks himself a great work, worthy the interposition of a great deity. More humble and I believe true to consider him created from animals.'[4]

The idea of all living things being linked through some sort of a transmutation process had been however, discussed across the continent since the early nineteenth century. In France the work of Jean-Baptiste Lamarck (1744–1829) envisioned that life was constantly generated in the form of the simplest creatures, which then strove towards complexity and perfection (i.e. humans) through a series of lower forms. As explained in Adrian J. Desmond's *The Politics of Evolution* (1989), in the early nineteenth century the work of Lamarck largely influenced the scientific, medical, and cultural debates all across Europe. In particular Lamarck's theory of evolution was discussed in Britain during the 1830s, especially in medical schools and radical circles, and contributed to form a new generation of scientists and doctors who saw in the work of the French naturalist an opportunity for a new understanding of the natural world.[5] The influence of Lamarck's theory of evolution, as we will see throughout this book, continued for the whole century and became central to the work of those — like Butler — whose aim was to think about evolution as something other than a purely mechanical process.

On Darwin's work, scholars have produced an extensive scholarship in recent years. The monumental *Darwinian Heritage* (1988) by David Kohn explored how Darwin's ideas were received and discussed in both England and abroad. Peter Bowler's *Evolution: The History of an Idea* (1989) and *The Eclipse of Darwinism* (1992) largely contributed to our current historical understanding of evolution in the nineteenth and twentieth centuries. In addition, *The Cambridge Companion to Darwin* (2003) explored several aspects of his work, including his engagement with religion and philosophy. Other studies — such as *The Reception of Charles Darwin in Europe* (2009), *The Cambridge Companion to the 'Origin of Species'* (2009), and Thomas F. Glick's *What about Darwin?* (2010) — have fruitfully discussed the influence that Darwin's work had on nineteenth-century science and culture.[6]

Darwin's work did not only attract the attention of historians. In recent years, the importance of Darwin's theory of evolution has also been largely discussed by literary scholars who explored the influence that the notion of natural selection had on the work of Victorian novelists. Gillian Beer, for example, in the now classic *Darwin's Plots* (1983) examined the place of Darwin's evolution in the writings of Charles Kingsley (1819–1875), George Eliot (1819–1880), and Thomas Hardy (1840–1928). Sally Shuttleworth's study of George Eliot has traces of references to evolution in her writing.[7] Likewise, George Levine's *Darwin and the Novelists* discusses the relationship between Darwinian evolution and Victorian fiction.[8] These are, of course, only a few examples of an extensive scholarship on the relationship between literature and science in the period but they still help us to understand how Darwin's ideas largely influenced the Victorian literary debate.

Darwin's *On the Origin of Species* was not, however, the only scientific text to attract the attention of the lay audience. Books about geology were also an object of popular fascination. For instance, in the 1860s, Charles Lyell's *Geological Evidences of the Antiquity of Man* (1863) became one of the most read and appreciated scientific books in England. The volume dealt with three scientific issues that had become prominent in the preceding decade: the age of the human race, the existence of ice ages, and Darwin's theory of evolution by natural selection.[9] In addition, Lyell's book contributed to a radical change of Victorian opinion about the study of our archaeological past and helped the establishment of prehistoric archaeology as a new scientific discipline.[10] Lyell's vivid writing, rich in analogies, influenced the public and helped the development of a popular scientific imagination. For example, both Jules Verne's *Journey to the Centre of the Earth* (1864) and Louis Figuier's 1867 second edition of *La Terre avant le déluge* drew largely on Lyell's ideas in creating their fictions.

In turn, across the nineteenth century, literature shared its language with science. Many examples are easily traceable: Lyell in his *Principles of Geology* (1830–33) extensively used the *Metamorphoses* of Ovid in explaining and describing his idea of proto-geology. In France, Claude Bernard (1813–78) in his *Experimental Medicine* (1865) cited exhaustively the work of Johann Wolfgang Goethe (1749–1832). Darwin was influenced by the work of William Paley (1743–1805) and Thomas Malthus (1766–1834).[11] As discussed by Beer in the article 'Darwin and Romanticism', the Victorian naturalist, during his expedition around the world on the Beagle, read and was influenced by the language, ideas, and myths of Romantic novels and poetical books including the poetical works of John Milton (1608–1674).[12] This is just an example among the many that show how science and literature in the nineteenth century shared not only a language but also ideas and a platform where to discuss them.

In addition to its exploration in literature, evolution also became the object of exhibitions, public lectures, and the topic of articles in various periodicals, which contributed to making natural history a key aspect of middle-class imagination. However, it is important to notice that the public representation and discussion of Darwin's theory was never neutral. Evolution was either discussed with devotion by some or harshly dismissed by others, often resulting in virulent quarrels. These disputes involved scientists, writers, and philosophers but also, in a variety of ways, the general public. It is among those debates that the Victorian writer Samuel Butler had a prominent and controversial role.

Samuel Butler and the Science of the Mind

Actively writing about evolution from the early 1860s to the end of the century, Samuel Butler was one of many Victorians who became fascinated by Darwin's theory of natural selection. In 1859, Butler had the chance to read Darwin's work for the first time, when he decided to leave England to move to New Zealand. On 20 December 1862, Butler anonymously published a dialogue on Darwin's *On the Origin of Species* in *The Press*, the local newspaper of the city of Christchurch.[13] Although written in a peculiar style, Butler's dialogue offered an accessible explanation of Darwin's hypothesis of evolution to New Zealand citizens. The narrative adopted by the British-born emigrant was a mix of satirical writing and scientific explanation combined with a deep philosophical analysis. As is well known, from the late 1870s onwards, Butler started a crusade against Darwin and his theory of evolution. However, in the early 1860s, Butler declared without hesitation: 'I was one of Mr. Darwin's many enthusiastic admirers, and wrote a philosophic dialogue (the most offensive form, except poetry and books of travel into supposed unknown countries, that even literature can assume) upon the Origin of Species.'[14]

Samuel Butler's writings on science have often been seen as just the attempt of an amateur to have a say in a debate that was far too big for his own understanding. However, in most of his books he tried to discuss and explore the work of Darwin while also promoting a 'new' account of evolution based on the work of the French naturalist Jean-Baptiste Lamarck. In 1885, in his notebooks, Butler summarized his work on evolution according to three major themes:

> 1. The identification of heredity and memory and the corollaries relating to sports, the reversion to remote ancestors, the phenomena of old age, the causes of the sterility of hybrids and the principles underlying longevity — all of which follow as a matter of course. This was *Life and Habit*. [1877.].
> 2. The re-introduction of teleology into organic life which, to me, seems hardly (if at all) less important than the Life and Habit theory. This was *Evolution Old and New*. [1879.].
> 3. An attempt to suggest an explanation of the physics of memory. I was alarmed by the suggestion and fathered it upon Professor Hering who never, that I can see, meant to say anything of the kind, but I forced my view on him, as it were, by taking hold of a sentence or two in his lecture, on *Memory as a Universal Function of Organised Matter* and thus connected memory with vibrations. This was *Unconscious Memory*. [1880.].[15]

As explained in the quotation above, Butler's idea of evolution involved more than a simple discussion or popularization of Darwin's theory of natural selection. For the Victorian writer, the theory of evolution became the starting point for a larger project, which involved biology, teleology, philosophy, and a critical discussion of the new physiological and psychological studies conducted on the brain in mainland Europe. Disagreeing with Darwin's theory of natural selection — which treated evolution as something purely mechanical — Butler tried to reintroduce design into the debate in the shape of a neo-Lamarckian philosophy of evolution. The methodology adopted by the author was to compare Darwinian science with other, similar and dissimilar, contemporary theories of evolution developed on the continent. Nonetheless, Butler also looked back to earlier generations of evolutionists, especially the works of the Comte de Buffon (1707–1788), Lamarck, and Erasmus Darwin (1731–1802). This was part of an attempt to write a history of evolution that aimed at explaining how Darwin's work had to be discussed in relation to previous theories.

However, Butler's work was not limited to a simple critical examination and discussion of the work of others. As a *post scriptum* of the previous note the writer declared:

> What I want to do now [1885] is to connect vibrations not only with memory but with the physical constitution of that body in which the memory resides,

> thus adopting Newland's law (sometimes called Mendelejeff's law) that there is only one substance, and that the characteristics of the vibrations going on within it at any given time will determine whether it will appear to us as (say) hydrogen, or sodium.[16]

Butler, in 1885, wanted to further explore the connection between memory and heredity by looking also at the biochemical structure of the human and animal body. In 1886, in his final book on evolution *Luck, or Cunning?*, he started discussing the process of vibration as the mechanism behind the accumulation of memory in the mind (brain) of the individual. Another extract from Butler's notebooks sheds some additional light on this process:

> I would make not only the mind, but the body of the organism to depend on the characteristics of the vibrations going on within it. The same vibrations which remind the chicken that it wants iron for its blood actually turn the pre-existing matter in the egg into the required material. According to this view the form and characteristics of the elements are as much the living expositions of certain vibrations — are as much our manner of perceiving that the vibrations going on in that part of the one universal substance are such and such — as the colour yellow is our perception that a substance is being struck by vibrations of light, so many to the second, or as the action of a man walking about is our mode of perceiving that such and such another combination of vibrations is, for the present, going on in the substance which, in consequence, has assumed the shape of the particular man. It is somewhere in this neighbourhood that I look for the connection between organic and inorganic.[17]

This is the starting point of this current investigation. Butler's writings on the science of the mind offers a unique opportunity for looking at a non-Darwinian theory of evolution, which is now partially forgotten. In this respect, to revive Butler's work on the mind means to explore the debate concerning Lamarckian evolution in England and continental Europe. Consequently, Butler's work needs to be discussed alongside other professionals and amateurs working in England and abroad. They include, but are not limited to: Ewald Hering (1834–1918) and his theory of memory as a form of heredity; the work of Théodule-Armand Ribot (1839–1913) on memory and experimental psychology; the research of George Romanes (1848–1894) on mental evolution, and instinct and intelligence; the writings of Herbert Spencer (1820–1903) on psychology and evolution; St George Jackson Mivart (1827–1900) and his criticism of Darwin's natural selection; and, finally, the work and ideas of Darwin himself.

In the past century, several biographical works on Butler's life and publications have been published, the most recent one by Peter Raby (*Samuel Butler: A Biography*) in 1991. In addition, several critical studies have also been published over the past two decades. These include Elinor Shaffer's *Erewhons of the Eye: Samuel Butler as Painter, Photographer, and Art Critic* (1988) and the edited collection *Samuel Butler, Victorian against the Grain* (2007) by James Paradis which provides a new and compelling discussion of the various aspects of Butler's work including art, music, and science. More recently, in 2015, David Gillott published a new monograph entitled *Samuel Butler against the Professionals,* which focused on Butler's artistic work and attitude toward specialism. What, however, has not yet been published is a comprehensive study focused solely on Butler's writings on science. In particular, Butler's ideas concerning evolution and the mind, in the form of a Lamarckian understanding of evolution, deserve to be discussed in more detail to unveil the importance of his contribution to the scientific and cultural debates of the time.

The complex relationship between Lamarckism and Darwinism will be a key aspect of this interpretation of Butler's work. In recent years, recognition of the role played by Lamarckism in the late nineteenth century has started to be increasingly relevant to our understanding of the history of nineteenth-century biology. Especially, the notion of 'inheritance of acquired characteristics' and the interpretation of evolution as a form of design made the theory of evolution advanced by Lamarck appealing for those, like Butler, who were looking away from the materialism of Darwin's work. Indeed, the work of Lamarck provided the perfect system for anyone who still wanted to explore evolution from a philosophical perspective.

Butler's attitude toward a more philosophical examination of evolution was reflected in both his novels and essays on science. Consequently, writing about Samuel Butler's contribution to Victorian speculative biology and culture requires an in-depth knowledge of both the literary and scientific debates (and their interconnections) of the late nineteenth century. Bernard Lightman, in the concluding chapter of the 2017 *Routledge Research Companion to Nineteenth-Century British Literature and Science*, explained that the study of literature and science has become, especially in recent years, an 'interdisciplinary exercise' in which historians of science and literary scholars share methodologies and ideas.[18] Lightman's argument is particularly relevant to the direction that this book will take in looking at Butler's use of evolution in both his fictional and non-fictional work.

There is, however, the need to recognize the role played by Butler's approach and the effect his work as a polymath had on the reception of his ideas. Butler was an elusive amateur who worked between two different eras of science. At the beginning of his career, science was still a gentlemanly and amateur matter. By the end of his career, it had become a field of expertise. Like many others of his generation, Butler found himself caught between two dissimilar moments of time and groups. Neither a professional nor solely a philosopher or a novelist, Butler worked among these groups sharing languages and methodologies, and becoming, in the end, an outsider. At the time of his death, he had no institutional or academic affiliation to secure his reputation, and his writings on science received little acknowledgement by the scientists of his generation. Nonetheless, Butler's science of the mind had a meaning in its own right, and it deserves to be discussed as a serious attempt to contribute to the late nineteenth-century evolutionary debate. In particular, Butler's writing on memory and the mind needs to be discussed within the context of experimental psychology as a new discipline that emerged across continental Europe in the 1870s.

In the late nineteenth and early twentieth centuries, psychology was viewed as a study of the phylogenetic development of mentality, as an experimental discipline aimed at exploring various functionalities of the brain including its response to external stimuli and at the study of the biological function of memory. However, there has been very little study on the birth of the discipline of psychology, and on the non-evolutionary strand of thinking about the mind as something that could be experimentally externalized and understood. Publications, such as Robert Richards' *Darwin and the Emergence of Evolutionary Theories of Mind and Behavior* (1989) and *The Meaning of Evolution* (1993) and Gregory Radick's *The Simian Tongue* (2007), show how historians have mainly restricted their thought within a Darwinian paradigm. However, Rick Rylance, in his recent book *Victorian Psychology and British Culture 1850–1880* (2000), recognized that Victorian psychology was considered as an open discourse with 'an audience crossing wide disciplinary interests. Economists, imaginative writers, philosophers, clerics, literary critics, policymakers, as well as biomedical scientists contributed to its formation'.[19]

Conversely, and paradoxically, the scholarship on evolution and psychology mentioned above has mostly overlooked a central evolutionary strand in its thinking and practice. There was a powerful current of thought, originating in France (e.g. Lamarck, Étienne Geoffroy Saint-Hilaire), and lasting from the nineteenth to the twentieth centuries, that treated memory as something that physically accrued in the body over generations. This memory was considered as something organically present in all living creatures providing an explanation for the complex phenomenon of heredity. These ideas were championed in Victorian Britain by Butler who developed them in essays, novels, and books. In his notebooks Butler wrote about this topic:

> Memory and heredity are the means of preserving experiences, of building them together, of uniting a mass of often confused detail into homogeneous and consistent mind and matter, but they do not originate. The increment in each generation, at the moment of its being an increment, has nothing to do with memory or heredity, it is due to the chances and changes of this mortal state.[20]

By putting together mind and matter, Butler unified philosophy with physiology, leaving a door open to the development of medical psychology. This conception of memory was often treated as a cul-de-sac by nineteenth-century psychologists, but in fact, these researchers felt an increasing pull of metaphysics in general, and embodied memory in particular, towards the end of their careers. The work of these people, and in particular that of Butler, is a trend that demands historical exploration rather than scientific effacement.[21]

Butler's main scientific books *Life and Habit* (1878) and *Unconscious Memory* (1881) presented a hypothesis based on the substantial overlap between the concept of memory and that of heredity, reintroducing causality into the process of evolution. These books anticipated the conclusion that the role of memory is central in the process of heredity, which became scientifically accepted only in the first decades of the twentieth century. The identities of memory and heredity adopted by Butler were justified by a phenomenon he called 'unconscious memory', which guaranteed the process of reproduction providing information from one generation to the other. The same idea was also developed (independently) in mainland Europe: in France by Ribot and in Germany by Hering.

The main claim of this research is to present Butler as one of the pioneers of the psychological study of the mind in England. This can only be achieved by looking at relationships, networks, debates, published and unpublished material (e.g. books, essays, notebooks, letters, and notes) produced by this eccentric Victorian during his lifetime. It is, therefore, important to investigate and recognize how Butler, instead of being purely a popularizer of Darwinian evolution, was able to discuss with both the professional and lay communities his ideas about memory, heredity, and the mind.

Synopsis of Chapters

The first chapter of this book explores the development of Butler's theory of memory as a form of heredity from the early 1860s to the turn of the twentieth century. Butler's idea of evolution was developed over the publication of four books, several articles, and essays between 1863 and 1890. Although never considered as a serious attempt to discuss evolution, Butler's various writings on science offer an insight into the complex debate concerning Darwinism and Lamarckism in the late nineteenth century. Indeed, Butler's theory of evolution was strongly in contrast with the materialistic approach suggested by Darwin's natural selection and proposed a return to a theory of evolution close to the one proposed in 1809 by Lamarck. Starting with a historical introduction, the aim of this chapter is to shed some light on Butler's idea of 'unconscious memory' and, in particular, to compare Butler's ideas with similar ones discussed in mainland Europe. In this latter respect, my claim is that Butler in his writing of science was not only a popularizer of the work of others. Instead, he was also trying to be an active protagonist of the debate, although only at a purely speculative level.

The second chapter looks at Butler's identity as a writer and explores his engagement with both literature and science in his fictional and non-fictional work. The overarching question of this chapter is: was Butler a novelist, a scientist, an artist, a philosopher, or a combination of these? In order to answer this question, the chapter provides a close reading of Butler's writing on science in three different sources: the series of articles 'God the Known and God the Unknown', his first book on science *Life and Habit*, and the novel *Erewhon*. The aim is to show how Butler blended science with literature, theology, and psychology. Particular attention is paid to Butler's writing on science in the novel *Erewhon*, the most successful book published during his lifetime. In contrast to the classic interpretation of Butler's novel as a work of satire, the chapter considers the novel partly as an attempt to write a speculative essay on evolution. Published in 1872, the novel sets out to popularize Darwin's natural selection. *Erewhon* produced a series of questions regarding Darwin's theory of evolution that crossed the boundaries between nineteenth-century literature and science. In *Erewhon*, Butler also introduced the concept of 'unconscious evolution', which would later become his main scientific hypothesis.

Chapter Three examines the reception of Butler's work in Britain. This section of the book sheds some light on the relationship between the Victorian writer and

Charles Darwin both before and after their famous quarrel. In doing so, the aim is to discuss and explore how Butler's relationship with Darwin changed over time and how this influenced the development of Butler's ideas and approach to science. The chapter discusses predominately two sources: it looks at the correspondence between the two Victorians from the early 1860s to Darwin's death in 1882 and the publication, content, and reception of one of Butler's most controversial books: *Evolution, Old and New*.

Chapter Four further explores Butler's writing on evolution and the mind, discussing, in particular, the place and significance of his ideas within the Victorian marketplace of science. The chapter also explores Butler's engagement and complex relationship with professional science. This is mostly done via a close examination of Butler's essay 'Remarks on George Romanes' Mental Evolution'. Here, the Victorian author attempted to show how Romanes' idea of mental evolution presented similarities with his theory of 'unconscious memory'. By looking at their different professional status and relationships with Darwin, the chapter investigates how these differences determined the professional and public reception of their theories.

The Conclusion, finally, discusses the rediscovery of Butler's literary and scientific ideas in the early twentieth century continental Europe. In France and Italy, particularly, Butler's work became the object of a cultural and scientific revival especially because of its use of Lamarckism. In France, Butler's work was discussed in the context of nineteenth-century French philosophy, while in Italy his work on the science of the mind was labelled as an example of neo-Lamarckism. The epilogue also serves the purpose to further stress how Butler's writings on the mind were ahead of their time. Indeed, as will be discussed, Butler's theory of memory and heredity was fully recognized only in the early twentieth century by Marcus Hartog (1851–1924) and Eugenio Rignano (1870–1930). Within the context of early twentieth-century debate on organic memory and neo-Lamarckism, they published several articles referring to Butler's and Hering's idea in the Italian periodical *Scientia*. These articles reveal how Butler's theory of memory and heredity had value but only within a more philosophical understanding of evolution. In particular, his status as a polymath working across different disciplines, which was considered his main weakness and posed a problem for the reception of his work in Victorian Britain, was finally recognized as one of his strengths.

Notes to the Introduction

1. Samuel Butler, *The Notebooks of Samuel Butler*, ed. by Henry Festing Jones (London: Fifield, 1912), p. 86.
2. George Levine, *Darwin and the Novelists: Patterns of Science in Victorian Fiction* (Chicago: University of Chicago Press, 1992); Gillian Beer, *Darwin's Plots: Evolutionary Narrative in Darwin, George Eliot and Nineteenth-Century Fiction* (Cambridge: Cambridge University Press, 2000); Adrian J. Desmond and James Richard Moore, *Darwin* (London: Norton & Co., 1992); George Levine, *Darwin the Writer* (Oxford: Oxford University Press, 2011).
3. See Steve Jones, *The Darwin Archipelago: The Naturalist's Career beyond Origin of Species* (London: Yale University Press, 2012), pp. 1–10 (p. 1).

4. See James Rachels, *Created from Animals: The Moral Implications of Darwinism* (Oxford: Oxford University Press, 1991), pp. 1–6 (p. 1).
5. See Adrian J. Desmond, *The Politics of Evolution: Morphology, Medicine, and Reform in Radical London* (Chicago: University of Chicago Press, 1989), pp. 276–335.
6. There is a rich scholarship on the reception of Darwin's theory in the Victorian cultural, social and scientific debate. See, for instance, Peter Bowler, *Evolution: The History of an Idea* (Berkley: University of California Press, 1989), pp. 237–46; Rebecca Stott, 'Darwin's Barnacles: Mid-Century Victorian Natural History and the Marine Grotesque', in *Transactions and Encounters: Science and Culture in the Nineteenth Century*, ed. by Roger Luckhurst and Josephine McDonagh (Manchester: Manchester University Press, 2002), pp. 151–82.
7. Sally Shuttleworth, *George Eliot and Nineteenth-Century Science: The Make-Believe of a Beginning* (Cambridge: Cambridge University Press, 1984), pp. 1–24.
8. Levine, *Darwin and the Novelists*, pp. 3–27 (p. 5).
9. A. Bowdoin Van Riper, *Men among the Mammoths* (Chicago: Chicago University Press, 1993), pp. 139–41.
10. Van Riper, *Men among the Mammoths*, p. 214.
11. See Beer, *Darwin's Plots*, p. 5.
12. See Gillian Beer, 'Darwin and Romanticism', *The Wordsworth Circle*, 41.1 (2010), 3–9 (p. 5).
13. Butler published 'Darwin on the Origin of Species: A Dialogue' in *The Press*, New Zealand, on 20 December 1862. The original 'Dialogue' is lost, but a reprinted version can be found in *The Press* of 8 June 1912 and in Samuel Butler, *A First Year in Canterbury Settlement: With Other Early Essays* (London: Fifield, 1914), pp. 155–64.
14. Samuel Butler, *Unconscious Memory* (London: Cape, 1920), p. 11.
15. Butler, *The Notebooks*, p. 66.
16. Butler, *The Notebooks*, p. 66.
17. Butler, *The Notebooks*, p. 67.
18. Bernard Lightman, 'Afterword', in *The Routledge Research Companion to Nineteenth-Century British Literature and Science*, ed. by John Holmes and Sharon Ruston (Oxford: Routledge, 2017), pp. 438–41 (p. 441).
19. Rick Rylance, *Victorian Psychology and British Culture 1850–1880* (Oxford: Oxford University Press, 2000), p. 7.
20. Butler, *The Notebooks*, p. 61.
21. Laura Otis' *Organic Memory* is an honourable exception to the general trend of this scholarship, but focuses more on the literary side of the question. See Laura Otis, *Organic Memory: History and the Body in the Late Nineteenth and Early Twentieth Centuries* (Lincoln: University of Nebraska Press, 1994).

CHAPTER 1

Butler and the Science of the Mind: A Pan-European Discussion

> To be is to think and to be thinkable. To live is to continue thinking and to remember having done so. Memory is to mind as viscosity is to protoplasm, it gives a tenacity to thought — a kind of *pied à terre* from which it can, and without which it could not, advance.
>
> Thought, in fact, and memory seem inseparable; no thought, no memory; and no memory, no thought. And, as conscious thought and conscious memory are functions one of another, so also are unconscious thought and unconscious memory. Memory is, as it were, the body of thought, and it is through memory that body and mind are linked together in rhythm or vibration; for body is such as it is by reason of the characteristics of the vibrations that are going on in it, and memory is only due to the fact that the vibrations are of such characteristics as to catch on to and be caught on to by other vibrations that flow into them from without — no catch, no memory.
>
> SAMUEL BUTLER, *The Notebooks*[1]

In *The Eclipse of Darwinism*, the historian of biology Peter Bowler, paraphrasing Julian Huxley's view on the modern evolutionary synthesis, placed the end of Darwinism between 1894 and the first decade of the twentieth century. However, strong criticism of Darwin's ideas, especially concerning a Lamarckian reading of evolution, had circulated in different forms across Europe since the 1860s. This was particularly evident within the debate concerning the mind and its psychological investigation.

Psychology was a branch of philosophy until the 1870s when it developed as an independent scientific discipline in Europe. Thus far, the development of psychology as an independent research discipline has often been discussed as a consequence of Darwin's hypothesis of evolution.[2] Although frequently linked with natural selection, psychology and, in particular, the study of the mind also developed alongside other scientific paths in both theory and practice across Europe. In respect of the practical aspect of the study of the human body and mind, the history of science offers an extensive literature. For instance, Laura Otis' publication *Müller's Lab* (2007) and Simon Schaffer's pamphlet *From Physics to Anthropology and Back Again* (1994) offer a historical view of the rise of nineteenth-century laboratories, although focusing their analysis on national cases. The work of Kurt Danziger (i.e. 'The Positivist Repudiation of Wundt' (1979) and *Constructing the Subject* (1990)) is an important point of reference for the historical scholarship interested in the development of psychology as an independent scientific discipline. Danziger offers a constructivist example of the intellectual history of psychological research from the nineteenth century (especially in Germany) to the emergence of contemporary psychology. The main aim of Danziger was to consider the psychological methodology as a social and cultural practice rather than as a simple matter of technique. However, in all of this literature, the aspect that is missing is a pan-European perspective and the consequent role of psychology in the cultural and philosophical debates.

As exemplified by Anson Rabinbach's *The Human Motor* (1992), the study of the historical, political, and cultural developments of nineteenth century science can only be conducted within a wider European perspective. Therefore, the rise of psychology and the history of its pioneers offer two of the best examples for understanding European science and culture between the 1860s and the turn of the twentieth century. The debate behind psychology, in particular, represents an important aspect of the late nineteenth-century European scientific scene. As a discipline involving the opinions of psychologists, physiologists, and philosophers, the study of the mind became an opportunity for understanding the merit of non-Darwinian interpretations of evolution in the British and continental scientific scenes.

The first aspect to discuss — in order to fully contextualize and understand the significance of Butler's work — is the heterogeneity of the European psychological debate. The psychological study of the mind did not have, especially at the outset, a proper scientific determination. It was, instead, a gathering of ideas, a rendezvous of psychologists, biologists, and philosophers with different backgrounds who were developing and promoting hypotheses sometimes in contrast to one another. Some general examples can indicate what comprised psychological research in the late nineteenth century. Scientists, especially in Germany and Italy, conducted physical experiments investigating the reaction of the human brain to external stimuli (e.g. light, colours, sounds) whilst inventing, testing, and producing new machinery and research tools. In Germany, the research of Wilhelm Maximilian Wundt (1832–1920), Ernst Mach (1838–1916), and Ewald Hering provided the basis for many twentieth-century discoveries in the field. They opened the first laboratories of experimental psychology in the world, marking psychology as an independent field of study. These laboratories also became examples of social and scientific formation through the establishment of new roles and practices. They contributed to the training of a new generation of physicists, psychologists, and physiologists who, thanks to these new research environments, worked collaboratively to develop new techniques and tools. The work of German experimental psychologists, in particular, was fundamental in creating a new platform for the study of the mind, which combined philosophical, psychological, and physiological ideas within one single framework.

In Britain, biologists, psychologists, and physiologists including, among many others, George Romanes and William Benjamin Carpenter (1813–1885) tested the physiology of the nervous system and also conducted experiments to lay the foundations for what Romanes called 'comparative psychology'. Romanes postulated a similarity between the cognitive processes and mechanisms of humans and other animals, declaring in *Animal Intelligence* (1878): 'there must be a psychological, no less than a physiological, continuity extending throughout the length and breadth of the animal kingdom.'[3] Philosophers also focused their attention on rethinking old metaphysical conceptions of the human mind (e.g. that of Immanuel Kant, John Locke, and David Hume) in response to the new discoveries made by scientists and psychologists. In Britain, the work of Herbert Spencer revolutionized the philosophical and psychological interpretation of evolution. In *Principles of Psychology* (1855) Spencer argued, that the human mind was subject to natural laws, which could be discovered within the framework of general biology. Spencer's work greatly influenced the research of James Sully (1842–1923) and James Ward (1843–1925), both of whom later contributed to the English psychological debate with the opening, in 1889, of the first two experimental psychology laboratories in the country: Sully at University College London and Ward at the University of Cambridge.

However, in contrast to Spencer's ideas, European philosophers tried alternative approaches. One of the main critics of Spencer's work was the Frenchman Théodule-Armand Ribot. In his research, Ribot used experimental and synthetic methods (strongly influenced by the work of his fellow countryman Claude Bernard) and brought together a large number of instances of inherited peculiarities. The case studies used by the French father of experimental psychology served the purpose to explain the complexity of the mind and the role it plays in memorizing biological information. Indeed, Ribot paid attention to the physical elements of mental life, ignoring all spiritual or non-material factors largely influencing the philosophical work of Henri-Louis Bergson (1859–1941). In contrast to Ribot's scientific approach, Bergson, in his *L'Evolution créatrice* (1907), forcefully proposed a return to a metaphysical notion of the mind. Bergson's *Creative Evolution,* additionally, attempted to think through the continuous creation of life, and explicitly pitted itself against Spencer's evolutionary philosophy.[4]

Finally, the psychological and physiological study of the mind also captured the attention of novelists and popularizers of science. They focused their works on this scientific and cultural revolution, embodying psychology in their literature and disseminating new theories and discoveries in novels, lectures, and articles along with new types of mass communication vehicles, especially periodicals. This is the context in which we should discuss Samuel Butler's idea of 'unconscious memory' in Britain, but also more broadly within a pan-European dimension. As discussed in the introduction, Butler actively wrote about evolution between the 1860s and 1890s. Working as a popularizer of European scientific ideas whilst trying to be an active protagonist of the British evolutionary debate, Butler is the perfect example for investigating the rise of Lamarckism (or perhaps the return to Lamarckism) and its influence on the development of the British psychological debate on evolution.

Memory, Evolution, and the Mind

On 2 December 1882, Samuel Butler delivered a lecture entitled 'On Memory as a Key to the Phenomena of Heredity' at the Working Men's College in London. In the lecture, Butler tried to explain to his fellow Londoners the importance of rethinking evolution in a Lamarckian way. In 1882, circumstances had led to Butler's voice not being given much credence within the British scientific community. Butler was, therefore, directing his attention to the general public with the aim of persuading the public to the revolutionary potential of his ideas concerning Lamarckism, memory, and heredity. Most significantly, the ideas that Butler championed were topics that were being widely discussed across Europe; however, in England where the faith in Darwin's theory of natural selection still remained strong, this 'new science' was initially met with reserve. Lamarck's work was well known in Britain since the 1830s and the publication of Charles Lyell's *Principles of Geology*. But after the publication of Darwin's *Origin of Species*, Lamarck's theory of evolution started to be ignored, primarily because of its emphasis on the will of the individual in the evolutionary process.

In his lecture, Butler sought to explain the role that memory played in heredity. He was aware that the question of exactly how inheritance occurred was still unanswered and he believed that an elaboration of Lamarck's concept of the inheritance of acquired characteristics would have provided a solution to the matter. However, to substantiate his claims, and perhaps to detract from his audience's doubt about either the scientific credibility of Lamarck's ideas, or of his own credentials to speak on this topic, Butler gave an account of his own research in the context of the work performed by European psychologists and neurologists. He insisted, in particular, in showing a similarity between his work and that of the German physiologist Ewald Hering. In the late nineteenth century, Hering was becoming well known across Europe for his research on heredity and memory, colour theory, and binocular vision. An extract from Butler's lecture states:

> We say it is a phenomenon of heredity that chickens should be laid as eggs in the first instance and clergymen born as babies, but, beyond the fact that we know heredity extremely well to look at and to do business with, we say that we know nothing about it. I have for some years maintained this to be a mistake and have urged, in company with Professor Hering, of Prague, and others, that the connection between memory and heredity is so close that there is no reason for regarding the two as generically different, though for convenience sake it may be well to specify them by different names.[5]

For Butler, the similarity between his ideas on memory and heredity and Hering's work was key to demonstrating the validity of his own theory. This is a central claim to make in looking at Butler's scientific work: while promoting his idea concerning the science of the mind, he was not simply popularizing European science. Instead, he tried to be an active protagonist in the debate by looking at evolution from both psychological and philosophical perspectives.

In his writing on science, Butler proposed a psychological elaboration of evolution (robustly enforced by Lamarck's philosophy), called 'unconscious memory'.

Butler's 'unconscious memory' was in stark contrast to the materialistic approach suggested by Darwin's natural selection and, as Butler argued, it had more scientific backing than the theory of pangenesis that Darwin had suggested as a potential physiological explanation of the mechanisms of heredity.

The reasons for Butler's dismissal of Darwin's ideas were both personal and methodological. On the personal side, as will be discussed in detail in chapter three, Butler had engaged with Darwin in a quarrel over different interpretations of evolution. The main consequence of the quarrel was the total isolation of Butler and his ideas from the British Darwinian community. At the heart of the quarrel lay a simple misunderstanding created by the forgotten acknowledgment of Butler's *Evolution, Old and New* (1879). Butler was convinced to find a reference to his book in the English translation of the biography of Erasmus Darwin by Ernst Krause (1839–1903), entitled *Life of Erasmus Darwin* (1879), which included a preface by Charles Darwin.[6] Several studies have explored the importance of the quarrel, focusing on both its public and private dimensions.[7] However, it is important to emphasize that the quarrel between Butler and Darwin was more than an *ad hominem* attack upon a forgotten citation.[8]

This was the case because for Butler evolution was a matter of Lamarckian designed memory. But for Darwin, there was little space for any blueprint in biology. As Janet Brown argues:

> pangenesis was the highly abstract notion that every tissue, cell and living part of an organism produced minute, unseen gemmules (or what he sometimes called granules or germs) which carried inheritable characteristics and were transmitted to the offspring via the reproductive process.[9]

In 1868, Darwin, in *The Variation of Animals and Plants under Domestication*, was careful to explain that each part of an organism produced only gemmules of itself and not of the organism as a whole.[10] Individual gemmules did not have the complete designed (biological) map of the whole creature.

Butler used the word 'unconscious' in relation to evolution for the first time in 1872 in his novel *Erewhon*. This term subsequently became the label the author adopted to describe his evolutionary idea. In his writing on science, Butler tried to rethink evolution and the notion of heredity as a mechanism which involved the reintroduction of causality. For the Victorian author, the only way to explain the hereditary processes was to show that, in nature, a substantial overlap between the concept of memory and heredity was key to solving the complex question concerning how specific characteristics were transferred from one generation to the next. Butler's idea of 'unconscious memory' rested on a process of biological reproduction and preservation of information from one generation to the other. This process was explained as a substantial chemical continuity between memory and heredity. This continuity was conceived in Butler's *Luck, or Cunning?* (1887) by employing a new interpretation of the Lamarckian concept of inheritance of acquired characteristics within a biochemical framework, as explained by Newlands' law. Butler, here, referred to the work of John Newlands (1839–1898), a British chemist who in the 1860s devised a periodic table of chemical elements.[11]

For Butler, it was important to think about memory as something organic that could be preserved within our body. Indeed, in his final book on science, *Luck, or Cunning, as the Main Means of Organic Modification?*, Butler made a clear statement:

> I shall perhaps best promote the acceptance of the two main points on which I have been insisting for some years past, I mean, the substantial identity between heredity and memory, and the reintroduction of design into organic development, by treating them as if they had something of that physical life with which they are so closely connected.[12]

As explained in the introduction, scholars have historically considered Samuel Butler as a novelist with an interest in science. Butler was a novelist, yes, but he also spent more than thirty years of his life fighting against the orthodoxy of Victorian science in order to explain to his fellow countrymen his vision of evolution. The place of Butler's science in contemporary Victorian studies is still very complex. If we take, as an example, the 2007 essay by Sally Shuttleworth, 'Evolutionary Psychology and *The Way of all Flesh*': it considers Butler's work predominately as a form of 'literature and science' without any relevant influence upon the scientific debate.[13] Shuttleworth explores the psychological meaning of Butler's most well-known novel looking at questions regarding personal identity and continuity of personality.[14] Her essay emphasizes the importance of looking at the theory presented in *Life and Habit* only through Butler's fiction, because fiction, according to Shuttleworth, is the only place where Butler's science can be taken seriously.[15] The complex relationship between literature and science in Butler's work will be discussed in chapter two, but for now it is important to stress how his writing on science developed organically through different channels.

In this respect, recent historical works are finally discussing the place and significance of Butler's writing in the Victorian scientific and cultural debate. The works of Paradis (2007) and Forsdyke (2006, 2009) have exemplified the importance of Butler's scientific view of evolution.[16] Moreover, the scientific work of Butler is now becoming increasingly recognized as forming a contribution to the Victorian marketplace of science although only in the form of popularization.[17] Bernard Lightman, in the 2007 essay 'A Conspiracy of One: Butler, Natural Theology, and Victorian Popularization', illustrates how the significance and place of Butler's work was largely limited to the popular sphere.[18] Lightman also explains that Butler's methodology, which did not involve any scientific experiment, 'represented a threat to the emerging scientific professionalization of Darwin's era'.[19]

The importance and significance of Butler's science is also partially explored in the introduction to Laura Otis' *Organic Memory* (1994), where Butler's idea of memory and heredity is briefly discussed alongside that of European psychologists and physiologists including Ewald Hering and Théodule-Armand Ribot. Otis' book invites the reader to think about the development of theories on heredity based on the idea of organic memory within a European context. In the volume, references to Butler's work are limited to the introduction, where Otis examines the historical development of the idea of organic memory in the late nineteenth and early twentieth centuries. Nonetheless, *Organic Memory* still makes two important points about the context in which we should discuss Butler's writings on science. First, Otis explains how any historical investigation of organic memory needs to be discussed moving beyond the work of just psychologists and physiologists by also including the ideas of writers, journalists, and philosophers who were all fascinated by this new discipline. Second, Otis also points out how the momentum for the development of ideas concerning the correlation between memory, heredity, and evolution was limited to a specific context and time period. Otis states: 'The association of heredity with memory in the nineteenth century grew out of a fascination with origins that manifested itself with the simultaneous rise of nationalism, philology, biology.'[20]

In sum, *Organic Memory* stresses the possibility of considering the consequences of the debate about organic memory beyond a scientific perspective. Otis divides her study into two different yet related levels. On the one hand, there is the scientific debate concerning the idea of organic memory and its developments in Germany, France, and England, which, as we will see, was largely discussed by Butler in his books on science. On the other hand, *Organic Memory* is a study focused on the use of organic memory in novels throughout the nineteenth and twentieth centuries. This is a less important aspect for the purpose of our current investigation, but it is still relevant to understanding how organic memory was never studied and discussed as just a scientific theory. So, when it comes to Butler, Otis' study shows exactly how individuals like him had a role in the development of nineteenth-century psychology. A similar argument is also marginally suggested by Daniel L. Schacter's history of psychology where Butler and Hering are presented as forgotten pioneers of the history of the discipline.[21]

In this light, there is, therefore, the need to fully re-examine and re-evaluate the role of Butler in the late Victorian scientific scene. Butler was not simply a popularizer of European science or a novelist interested in evolution. Instead, he tried to be an active protagonist in the debate by looking at evolution from a psychological and philosophical perspective. In particular, Butler's scientific writings need to be taken seriously as they can also help us to understand the pan-European dimension of the debate about evolution and the mind. Influenced by Lamarck, Hering, and Ribot, Butler sought to bring the theories proposed by these three thinkers, which were influential and widely accepted in Europe, to Britain. However, his complex relationship with Darwin and other Victorian scientists undermined his attempts to do so.

It follows that the only way to understand Butler's writings on science and the mind is to look at them in the context of nineteenth-century European experimental psychology. Only in this way we will be able to fully comprehend the importance of the mind in the hereditary process. In particular, when it comes to Butler, his writings on science need to be discussed within a pan-European context, which involves the rise of experimental psychology as one of the necessary conditions for understanding the complexity of his theory of 'unconscious memory'.

Butler's Scientific Writings: The European Connections

Samuel Butler's work on evolution was based on a large critical review of the main nineteenth-century scientific texts. Between 1863 and 1890 Butler read, translated, and popularized a considerable amount of English, French, and German evolutionary literature. Both historians of science and literary scholars have recognized how, in the nineteenth century, there was a continuity between the work, language, and even ideas of novelists and men of science.[22] However, it is important to stress that in the Victorian period there was also a clear distinction between practitioners and popularizers.[23] As Lightman has explained in his seminal work *Victorian Popularizers of Science*, the Victorian popularizer was often not a practitioner and their work was generally focused on the cultural understanding of one particular scientific idea.[24] Paul White's book on Thomas Huxley (1825–1895) also looks at the establishment of professionalism in the 1870s stressing the need to distinguish the creation of scientific knowledge from its popularization.[25] Indeed, White investigates the role of Huxley in the passage from museum-based learning to a laboratory-based study of nature.[26] White explains that '[i]n his science columns for the Westminster, Huxley drew sharp distinctions between works of genius, which produced new knowledge, works of popularization, which disseminated knowledge, and works of popular delusion or superstition, which debased knowledge'.[27] For Huxley, therefore, the establishment of scientific laboratories created a distance between amateurs and professionals both in terms of working place but also, more importantly, status. There is, consequently — as suggested by Lightman — a distinction to make between the Victorian practitioner who produced scientific knowledge conducting experiments in a specific laboratory, using proper tools and publishing in well-recognized scientific journals, and the popularizer whose main job was to entertain the masses by talking about science.[28]

In *Life and Habit* (1878), Butler went against this new tradition and made a controversial statement about the production of scientific knowledge: 'I say that the term "scientific" should be applied (only that they would not like it) to the nice sensible people who know what's what rather than to the discovering class.'[29] Therefore, for Butler, as will be discussed in detail later on, making science was not simply a question of conducting experiments or collecting specimens in remote locations. Instead, it was possible to produce new scientific knowledge simply via knowing and reflecting on the ideas of others.

Samuel Butler made this way of producing scientific literature his personal writing style. He read, commented on, and critiqued Darwinian and non-Darwinian ideas and tried, especially in *Evolution, Old and New*, to link the present Darwinian science with the Lamarckism of the past. Butler endeavoured to show how certain ideas proposed by Darwin owed a deep debt to the work of the previous generation of naturalists, especially Lamarck. However, he also tried to look at how evolution was currently discussed across Europe with a particular focus on the notion of memory. Indeed, Butler's writings on science aimed to link a historical examination of the past with a speculative discussion of the current European science. Although this approach was not too common in England, in Germany speculative discussions of evolution, especially in the form of a philosophical examination of nature, were still central to the intellectual debate of the time.

One such example of this tradition is the work of Eduard von Hartmann (1842–1906) who, in 1869, published his *Philosophie des Unbewussten*. The volume, which obtained modest success in Germany, looked at Darwin's ideas concerning evolution in order to philosophically discuss concepts such as reason, will, design, and the mind. Von Hartmann's idea was well received by the philosophical community. This is confirmed by general reviews that defined his work as 'phenomenal', especially in its homeland.[30] However, although recognizing von Hartmann's knowledge of the physiological and evolutionary debate of his time, von Hartmann's work was mostly classified as philosophical investigation with little impact on the actual scientific debate.[31] This conclusion is evident in an article by James Sully published in 1876 in the *Westminster Review*. Indeed, Sully recognized the importance of von Hartmann's work, but, also its limits:

> Subtract this questionable factor — the unconscious from Hartmann's 'Biology and Psychology,' and the chapters remain pleasant and instructive reading. But with the third part of his work — the Metaphysic of the Unconscious — our feet are clogged at every step. We are encircled by the merest play of words, the most unsatisfactory demonstrations, and most inconsistent inferences. The theory of final causes has been hitherto employed to show the wisdom of the world; with our Pessimist philosopher it shows nothing but its irrationality and misery. Consciousness has been generally supposed to be the condition of all happiness and interest in life; here it simply awakens us to misery, and the lower an animal lies in the scale of conscious life, the better and the pleasanter.[32]

Like that of Butler, the philosophical work of Eduard von Hartmann struggled to impress the scientific community due to his lack of solid evidence and the complex philosophical language.

In 1878, with the publication of *Life and Habit*, Butler recognized that the first naturalist to identify a link between memory and heredity was Jean-Baptiste Lamarck in his work, *Philosophie zoologique* (i.e. the concept of 'inheritance of acquired characteristics'). Butler's use of Lamarck was due to his desire to propose 'the re-introduction of teleology into organic life'.[33] However, Butler's intention in resurrecting Lamarck's philosophy was not just instrumental to his criticism of Darwin's natural selection. Butler's aim was also to complement Darwinism with Lamarckism via a historical examination of the two ideas.

Lamarck's *Philosophie zoologique* (1809) proposed a hypothesis of evolution (called *transformisme*) where the idea of transformation implied a designed evolution of living species. Lamarck's philosophy was primarily progressive, although it did involve some divergences.[34] Butler accepted Lamarck's idea that in nature there is no extinction, and that evolution determines the passage from simple to complex forms of life based on a continuous reproduction of an 'inheritance of acquired characteristics'. Lamarckian evolution was based on a process of adaptation of the organism to its environment. This adaptation was explained by the naturalist via a process of use and disuse of certain parts of the body or characteristics.[35]

In *Philosophie zoologique*, Lamarck located the source of vital stimulation within the nervous system. Following the eighteenth-century physiological tradition, Lamarck considered the nervous fluids as the principle link existing between living things and the environment.[36] In the *Histoire naturelle des animaux sans vertèbres* (1815), Lamarck even identified an organic component, a fluid, to determine the functions of the living body.[37]

It is worth repeating, how in *The Politics of Evolution*, Adrian J. Desmond explained that Lamarck's work influenced the early nineteenth-century scientific debate in England (but also in wider Europe) and further shaped the medical and biological background of the next generation of scientists. However, it is also important to observe that after the publication of the *Origin of Species*, Lamarck's work started to be dismissed by many in Britain. Butler's scientific work and desire to resurrect Lamarckian ideas, therefore, becomes a primary example with which to understand the importance of Lamarck's idea of 'inheritance' in Europe. In particular, in *Unconscious Memory* and *Luck, or Cunning?*, Butler explained how Lamarck's hypothesis of inheritance developed in both physiology and philosophy across the continent. Butler recognized traces of Lamarckism in the research of Ewald Hering and Ernst Haeckel (1834–1919), among others, in Germany, Théodule-Armand Ribot in France, and William Benjamin Carpenter, Herbert Spencer and George Romanes in England.

Haeckel's vision of evolution was made of a new synthesis of Darwin's ideas with the German tradition of *Naturphilosophie* (referring in particular to Johann Wolfgang Goethe and Gottfried Wilhelm Leibniz) and the progressive evolutionism of Lamarck. In *Generelle Morphologie* (1866) Haeckel's biogenetic law proposed Lamarck's hypothesis of *transformisme* again, but in a different shape. The German biologist, although being an energetic defender of Darwinism, attempted to demonstrate that all multicellular organisms evolved from a common ancestor.[38] Haeckel's physiological research considered the process of evolution as being directly related to the organism, and suggested a strong link between the work of Darwin (especially the *Origin of Species*) and Lamarck's idea of inheritance. However, unlike Darwin, who concentrated on the materialistic side of evolution, Haeckel considered the topic of evolution from a metaphysical perspective.

Haeckel's idea was to develop an evolutionary metaphysics in order to rethink evolution as a monistic, unified account of life. This approach, in reshaping Darwin's work, sought to use psychology and the study of the mind to develop a new framework for understanding evolution. Consequently, evolution, for Haeckel, involved and encompassed not only nature but also the evolution of the psyche. However, Haeckel proposed only a theoretical interpretation of some of the key ideas that would be central to the experimental debate of the 1870s. After the work of Haeckel, the experimental study of 'inheritance' was undertaken by Ewald Hering.

Hering is fundamental for Samuel Butler's popularization of Lamarckian ideas. From 1880 and the publication of *Unconscious Memory*, Butler cited and discussed Hering's work in all of his scientific books or essays. Educated as a physicist in Leipzig, Hering subsequently worked as a physiologist at the University of Vienna between 1865 and 1870, in Prague between 1870 and 1895, and in Leipzig between

1895 and 1908.³⁹ As a physiologist, Hering became known in Europe largely due to his research into colour vision and spatial perception. Nonetheless, he was also the first, in Germany, to suggest that the idea of organic memory could be adopted to explain the hereditary process, and to conduct experiments on it.⁴⁰

Hering worked at the heart of a dynamic debate and conducted different types of research. In Vienna, he challenged physiologist Hermann von Helmholtz's (1821–1894) colour vision theory. At the same university, he conducted research on respiration and, in 1868, with psychoanalyst Josef Breuer (1842–1925), demonstrated the role of the vagus nerve in the regulation of breathing. However, Hering's work was also influenced by Goethe's scientific theory of colours and the philosophies of Kant, Friedrich Wilhelm Joseph Schelling (1755–1854) and Johann Gottlieb Fichte (1762–1814), alongside the work of physiologists like Johannes Müller (1801–1858) and Haeckel. As a result, his approach to the subject was partly philosophical and partly scientific. In one of his latest works on nerve activity (1900), Hering combined a rigorous discussion of the physiological study of the nervous system with a speculative discussion on questions concerning monism and the life of the individual.⁴¹ This trend of combining scientific and philosophical lines of inquiries is confirmed by the reception of his work. For instance, the review of Hering's book *On Memory and the Specific Energies of the Nervous System* (1895), published by *The Monist*, stressed the importance of Hering's approach by saying: 'more of such work would have been a boon to the semi-scientific world.'⁴²

In 1870, Hering presented a lecture entitled 'Über das Gedächtnis als eine allgemeine Funktion der organisierten Materie' ('On Memory as a Universal Function of Organized Matter') at the University of Prague. The lecture rapidly became one of the most frequently quoted texts in the field. It gave rise to a series of translations and was largely used among European physiologists. In Britain, the first reference to Hering and the notion of memory and heredity was published by Ray E. Lankester (1847–1929) in *Nature* in 1876 under the title of: 'Perigenesis v. Pangenesis: Haeckel's New Theory of Heredity'. Lankester briefly mentioned the name of Hering but without providing a full account of his theory. The first full account of Hering's work was, then, provided by Butler in 1880 in *Unconscious Memory* where the paper was translated and published as an integral part of the book.⁴³ I will return to the significance of Butler's translation of Hering's lecture later when discussing his remarks on George Romanes' *Mental Evolution in Animals*.

For now, what is important is that Hering's paper identified memory as a fundamental reproductive capability of living matter. The main scientific hypothesis enclosed in Hering's lecture was the necessity to link materialistic science (physiology) with the philosophy of the mind (psychology). Hering's study focused on memory linked to the body, moving it into the realms of physiological processes. It involved scientific concepts such as reproduction, conservation changes, and memory as hereditary, but also various philosophical problems. A quote from Hering's lecture exemplifies this point:

> The word 'memory' is often understood as though it meant nothing more than

> our faculty of intentionally reproducing ideas or series of ideas. But when the figures and events of bygone days rise up again unbidden in our minds, is not this also an act of recollection or memory? We have a perfect right to extend our conception of memory so as to make it embrace involuntary reproductions, of sensations, ideas, perceptions, and efforts; but we find, on having done so, that we have so far enlarged her boundaries that she proves to be an ultimate and original power, the source, and at the same time the unifying bond, of our whole conscious life.[44]

Hering located the origin of human memory, and that of animals and plants, in the reflexes and instincts of primitive ancestors.[45] Consequently, memory was incorporated into the research of the human nervous system; memory became, in Hering's work, part of a new physiological interpretation of the human body in which the brain and the nervous system were considered as the base for the new medicine of the human body. Memory was also the key to explaining the significance of heredity as a continuity between generations without a drastic denial of its philosophical importance. Hering seems to focus particularly on this point at the end of the lecture where he declares without any further doubts:

> The most sublime ideas, though never so immortalised in speech or letters, are yet nothing for heads that are out of harmony with them; they must be not only heard, but reproduced; and both speech and writing would be in vain were there not an inheritance of inward and outward brain development, growing in correspondence with the inheritance of ideas that are handed down from age to age, and did not an enhanced capacity for their reproduction on the part of each succeeding generation accompany the thoughts that have been preserved in writing. Man's conscious memory comes to an end at death, but the unconscious memory of Nature is true and ineradicable: whoever succeeds in stamping upon her the impress of his work, she will remember him to the end of time.[46]

In this period, Lamarckism was also discussed in France. In his scientific writing, Butler cited the works of important French Lamarckian biologists such as Yves Delage (1854–1920), Félix Le Dantec (1869–1917), Jean-Louis de Lanessan (1843–1919), and Jean Louis Armand de Quatrefages de Bréau (1810–1892). They all conducted research very close to Lamarckian ideas. They considered Darwin's work as a simple development of Lamarck's evolutionist paradigm.[47] Even more controversially, de Lanessan in his 1883 book *Le Transformisme* negated any originality to Darwin's work.

In *Life and Habit*, in particular, Butler largely refers to the work of another French scientist: Ribot. After his appointment as Professor at the Collège de France in 1880s, Ribot opened the first laboratory of experimental psychology in the country. He defined memory and heredity along the same lines as Hering. However, in contrast to the German tradition, which was based mostly on empirical research, Ribot divided his time between empirical research and promoting psychology to both scientists and the general public. In addition, Ribot was also involved in several philosophical debates concerning, for example, the nature of psychological research or the role of positivism in nineteenth-century science.[48]

Ribot's main scholarly interest was to embed 'memory' into physiological research whilst remaining aware of its philosophical roots. It is, therefore, interesting to note the research approach adopted by the French psychologist. Although Ribot was trained in philosophy, he practiced clinical and experimental psychology from 1873 to 1885. As suggested by Vincent Gullin, Ribot did not only introduce experimental psychology in France but also reshaped the study of natural science in relation to 'l'anatomie, la physiologie, la pathologie mentale, l'histoire, l'anthropologie'.[49] Ribot's research was therefore received with interest by the scientific community. Nonetheless, it also attracted the interest of philosophers like Henri Bergson.[50] Otis explains that between the 1880s and 1890s, the neurological journal *Brain* frequently cited Ribot's research and his *Le Maladies de la mémoire* became the most quoted neurological publication of the late nineteenth century.[51] Therefore, the significance of Ribot's research went far beyond the development of experimental psychology.

Memory, as defined by Ribot in the introduction of *Les Maladies de la mémoire*, was '[p]ar essence, un fait biologique; par accident un fait psychologique'[52] and made sense only when merged with heredity, instinct, and habit.[53] Indeed, for the French psychologist memory and heredity are intrinsically the same. Ribot's hypothesis criticized the orthodoxy of biology which gave precedent to conscious memory and cut it off from the domains of the unconscious (i.e. memory as a biological phenomenon).

Like Hering, Ribot recognized the potential of Lamarck's idea of 'inheritance' and its role in evolution. For Ribot, memory could only be described using a new scientific terminology and it was not made of an 'indefinable' metaphysical substance. It was, instead, a biochemical composition which leaves physical traces and residues. In this way, memory became subject to a process of accumulation ('le capital accumulé') which, citing the work of Henry Maudsley (1835–1918) and Joseph Delboeuf (1831–1896), Ribot called 'les mouvements moléculaires' ('molecular vibration').[54] Memory, for Ribot, required a dynamic association that, through repetition, established a stable, primitive anatomical connection.[55]

In England, Butler recognized and discussed Lamarckian ideas in the work of St. George Mivart, William Benjamin Carpenter, Herbert Spencer, and George Romanes. In *Life and Habit*, Butler explored in detail Carpenter's *Principles of Mental Physiology* (1874). Butler suggested that Carpenter produced one of the first physiological interpretations of the role of the mind in the economy of the body, largely influencing the subsequent medical and physiological debates. Butler recognized only one problem in Carpenter's work: 'The only issue between myself and Dr. Carpenter would appear to be, that Dr. Carpenter, himself an acknowledged leader in the scientific world, restricts the term 'scientific' to the people who know that they know.'[56]

In *Unconscious Memory*, Butler discussed Spencer's contribution to the scientific and philosophical debate. Butler recognized Spencer's scientific writing as the cornerstone for a new interpretation of evolution in between Darwinism and Lamarckism. Butler appreciated that Spencer's work (which was also translated into

French by Ribot) was purely theoretical and directed to a specialized philosophical audience. Butler's opinion of Spencer's work changed over the years. In 1884, Butler partially dismissed Spencer's work declaring:

> no writer that I know of except Professor Hering of Prague, [...] has shown a comprehension of the fact that these expressions are unexplained so long as 'heredity,' whereby they explain them, is unexplained; and none of them sees the importance of emphasizing Memory, and making it as it were the keystone of the system.[57]

However, in 1889, in the essay 'The Deadlock in Darwinism', Butler recognized that '[t]he Lamarckian system has all along been maintained by Mr. Herbert Spencer'.[58] Butler was not able to see 'any important difference in the main position taken by him and Lamarck'.[59]

Finally, mention should be made of the work of George Romanes. In 1881, Romanes, while taking a position in the Darwin-Butler quarrel in defence of the naturalist, publicly rejected Butler's hypothesis of unconscious evolution in his review of *Unconscious Memory* published in *Nature*. Romanes insisted on showing how Butler's ideas did not have any scientific value because he was not a professional. Romanes wrote on Butler's incompetence: 'To this arena [science], however, he is in no way adapted, either by mental status or mental equipment.'[60] In response, in 1884, Butler published a short essay entitled *Remarks on George Romanes' Mental Evolution in Animals*. The aim of the essay, which will be discussed in detail later on, was to show that Romanes in his *Mental Evolution in Animals* (1883) used Lamarckism in a manner similar to *Life and Habit*.[61]

Samuel Butler and the Idea of 'Unconscious Memory'

In Butler's work, the word 'unconscious' describes mechanical actions of a living body including breathing, blood circulation, and embryological reproduction, but also 'actions which we have acquired with difficulty and now perform almost unconsciously' such as 'playing a difficult piece of music, reading, talking, walking'.[62] All of those actions were guaranteed by the presence of a (biological) memory in the human body. Memory, for Butler, was the key aspect of the hereditary process because it could be physically reproduced. This was explained clearly in his notebooks:

> There is the reproduction of an idea which has been produced once already, and there is the reproduction of a living form which has been produced once already. The first reproduction is certainly an effort of memory. It should not therefore surprise us if the second reproduction should turn out to be an effort of memory also. Indeed all forms of reproduction that we can follow are based directly or indirectly upon memory.[63]

In order to understand the significance of Butler's theory of memory and heredity it is important to trace its development. At the beginning of his career, Butler referred to 'unconscious memory' in his novels and short essays. Between 1863 and 1878, Butler engaged with the debate of evolution from a purely philosophical perspective.

This interest is particularly evident in the periodical articles 'Darwin among the Machines' (1863), 'Lucubratio Ebria' (1865), the novel *Erewhon* (1872) and *Life and Habit*. In the 1860s, Butler published two short philosophical articles in the New Zealand periodical *The Press*: 'Darwin among the Machines' and 'Lucubratio Ebria'. In these articles, Butler attempted to explain Darwin's theory of natural selection in terms of the evolution of machines.[64] Butler also tried to question the role played by mechanical tools such as notebooks and umbrellas in human evolution. In these articles, he offered an anthropological reading of Darwin's natural selection in a language full of sarcasm and improbable analogies. Butler's intention was not only to teach or question evolution but, importantly, to also make this new scientific theory accessible to a New Zealand audience.

In *Erewhon*, as will be shown in more detail in the next chapter, Butler expanded upon the idea of 'unconscious memory', by producing a twenty-five-page manifesto in which he discussed the difference between conscious and unconscious actions whilst describing organic and inorganic evolution. As Roger Robinson has explained, in the three chapters entitled 'The Book of the Machines' Butler wrote his personal eulogy to Darwin's *Origin of Species* taking evolution to its paradoxical extremes.[65] The main merit of *Erewhon*, in advancing Butler's position about psychological evolution, is to finally present the potential of 'unconscious memory'. In the novel, 'unconscious memory' is the 'medium' which permits the preservation of life and makes the generation of mechanical life possible.

Six years after the publication of *Erewhon*, Butler returned to 'unconscious memory' by proposing his idea in a different manner. After temporarily leaving his occupation as a novelist, between 1876 and 1877, Butler produced his first speculative book about unconscious evolution: *Life and Habit*. He summarized *Life and Habit*'s theme as: 'The identification of heredity and memory'.[66] Although the book was presented as a 'scientific publication', by discussing many of the topics that were in vogue during the period, Butler's critical approach can still be considered an example of 'natural philosophy'. It is important to be clear regarding the philosophical nature of the text because this can explain why it was overlooked by the scientific community. *Life and Habit* was based on a critical examination of Darwin's natural selection mediated with Lamarck's hypothesis of 'inheritance of acquired characteristics'. However, it also engaged with philosophical topics including metaphysical and epistemological questions. *Life and Habit* aimed to be what Robert Chambers' *Vestiges of the Natural History of Creation* had been in the 1840s. Butler tried to present an argument that was engaging for philosophers, scientists, and the public audience. However, his book also aimed to be, as declared by the writer in the text, 'a valuable adjunct to Darwinism'.[67] The 'valuable adjunct' was, indeed, Lamarck's philosophy of evolution. In *Life and Habit*, Lamarck played a key role. Butler himself made this clear declaring in *Luck, or Cunning?*: 'to Lamarck, therefore, I naturally turned, and soon saw that the theory on which I had been insisting in "Life and Habit" was in reality an easy corollary on his system.'[68]

In *Life and Habit*, Butler also discussed Mivart's *Genesis of Species* (1871), Carpenter's *Mental Physiology* (1874), and Ribot's *L'Hérédité: Étude psychologique* (1873), especially in relation to the research of Henry Maudsley. Additionally, it was in this work that he provided accounts of Aristotle, Socrates, Plato, Marcus Aurelius, and St Paul in order to establish a strong link between Victorian science and an older metaphysics. Butler's intention was to offer to readers an accessible way to understand everything regarding evolution and not just present the results of scientific research.

The philosophical nature of the book was also illustrated by the examples and terminology Butler used. In *Life and Habit*, Butler explained — citing and discussing large portions of Ribot's work — that humans have two different types of memory: 'intelligence' and 'instinct'. 'Intelligence' is the mode of memory acquired through learning and habits. 'Instinct', by contrast, is a type of memory which exists in our cells and connects any living creature with its own ancestors. In explaining this difference Butler directly cited Ribot's *L'Hérédité*: '"Whereas intelligence is developed slowly by accumulated experience, instinct is perfect from the first" ('Heredity', p. 14).'[69] In *Life and Habit*, Butler's theory of heredity was very close to Ribot's although with one notable difference: if, for Ribot, memory could only be understood in mechanical terms or as biological accumulation ('le capital accumulé'), for Butler there was still something missing. He wrote:

> Obviously the memory of a habit or experience will not commonly be transmitted to offspring in that perfection which is called 'instinct,' till the habit or experience has been repeated in several generations with more or less uniformity; for otherwise the impression made will not be strong enough to endure through the busy and difficult task of reproduction.[70]

Memory, for Butler, was something more than a simple mechanical ability; it was the element that links the physical structure of the brain with its metaphysical nature. For Butler, memory and body were linked together, as well as memory and heredity. Paraphrasing *Life and Habit*, memory and heredity are the means of preserving experiences and carrying them to the next generation.

Life and Habit did not receive enough attention from Victorian readers, and Butler's theory was dismissed as an example of a philosophy of life lacking any serious scientific acumen. In response to this criticism Butler, in 1879, published *Evolution, Old and New*, where he attempted to trace the development of the idea of evolution before Charles Darwin. Although not enlarging upon Butler's theorization of unconscious evolution, the book provides an overview of Lamarck's idea of 'inheritance' and its influence on the pre- and post-Darwinian British debate. Butler himself declared in *Luck, or Cunning?*: 'I wrote "Life and Habit" to show that our mental and bodily acquisitions were mainly stores of memory: I wrote "Evolution Old and New" to add that the memory must be a mindful and designing memory.'[71] This explains the secondary aim of *Evolution, Old and New*, which was to present memory as something between 'matter' and 'metaphysics', linking the work of Darwin with Lamarckism and highlighting their differences and similarities.

In 1880, Butler tried to propose the idea of memory as heredity for the second time. With the publication of *Unconscious Memory*, Butler returned to the theory that the scientific community had as yet found unconvincing. It was in this articulation of his conception of the role of memory in evolution that Butler drew from Hering's

1870 lecture 'Über das Gedächtnis als eine allgemeine Funktion der organisierten Materie'. It can be argued that the book is nothing more than a discussion of Hering's work used by Butler as a justification of his own idea. This is because Hering's writing anticipated Butler's theory and used a language and a methodology far better suited to the persuasion of the scientific community. Therefore, Butler decided to dedicate a large part of *Unconscious Memory* to Hering's work, providing an English translation of the lecture. Speaking of this decision, Butler wrote (referring to Hering and himself):

> If two men so placed, after years of reflection, arrive independently of one another at an identical conclusion as regards the manner in which this machinery must have been invented and perfected, it is natural that each should take a deep interest in the arguments of the other, and be anxious to put them forward with the utmost possible prominence.[72]

Unconscious Memory presents two main differences to Butler's previous works. Firstly, Butler partially accepted Hering's theory of memory as a form of molecular vibration. The vibration theory was defined as a series of chemical changes that occur in a substance called 'protoplasm' through repetition.[73] The word 'protoplasm' comes from the Greek words *protos* (first) and *plasma* (anything formed). Protoplasm was introduced to the scientific language in 1846 by the German botanist Hugo von Mohl (1805–1872). It was defined as the 'tough, slimy, granular, semi-fluid' substance within plant cells but different from the cell wall, nucleus, and sap within the vacuole. In *Unconscious Memory*, Butler explained that protoplasm 'may be, and perhaps is, the *most* living part of an organism, as the most capable of retaining vibrations'.[74]

The concept of protoplasm became very popular among British biologists. In 1869 Huxley, in a famous pamphlet, defined protoplasm as the 'physical basis of life'.[75] In 1879, George J. Allaman wrote in *The Popular Science Monthly*:

> Protoplasm lies at the base of every vital phenomenon. It is, as Huxley has expressed it, 'the physical basis of life;' wherever there is life from its lowest to its highest manifestation there is protoplasm; wherever there is protoplasm there is life.[76]

However, the science of protoplasm was not certain or precise. Butler, in particular, was not fully convinced by this new theory. He wrote in *Luck, or Cunning?*:

> Science has not, I believe, settled all the components of protoplasm, but this is neither here nor there; she has settled what it is in great part, and there is no trusting her not to settle the rest at any moment, even if she has not already done so.[77]

The first full chemical description of protoplasm was only published in 1938, by the American chemist E. Newton Harvey (1887–1959) in the article: 'Some Physical Properties of Protoplasm'. Harvey described protoplasm as: 'an albuminous substance containing carbon, hydrogen, oxygen and nitrogen in extremely complex molecular combination and capable under proper condition of manifesting certain vital phenomena'.[78]

In *Unconscious Memory*, Butler also expressed some partial doubts on the vibration theory:

> I am not committed to the vibration theory of memory, though inclined to accept it on a *primâ facie* view. All I am committed to is, that if memory is due to persistence of vibrations, so is heredity; and if memory is not so due, then no more is heredity.[79]

In saying this, Butler did not reject his Lamarckian view of heredity proposed in 1878. Instead, he suggested that whilst he knew nothing about the vibration theory when he wrote *Life and Habit*, this new biological advancement did not affect his theory of memory and heredity.

Unconscious Memory presented another novelty. In *Life and Habit*, Butler defined himself as a member of the general public with the intention of explaining evolution to a popular audience. In 1880, he left the naive spirit of the previous books and placed himself next to Hering, whilst still highlighting his status as a non-practitioner of science. He wrote:

> Professor Hering and I, to use a metaphor of his own, are as men who have observed the action of living beings upon the stage of the world, he from the point of view at once of a spectator and of one who has free access to much of what goes on behind the scenes, I from that of a spectator only, with none but the vaguest notion of the actual manner in which the stage machinery is worked.[80]

In his final and most polemical book, *Luck, or Cunning?*, Butler again proposed the Lamarckian mechanism of 'unconscious memory'. He tried to show how Lamarck's theory of memory and heredity was implicit in much of the teachings of Spencer, Romanes, and other leading biologists — although hidden by the Darwinian shadow.

Luck, or Cunning? did not present any significant advancement of Butler's theory of memory as heredity proposed in *Unconscious Memory*. However, Butler returned to the idea of protoplasm and memory. In particular, he accepted a more marked development in the vibration hypothesis of memory given by Hering and only adopted with reserve in *Unconscious Memory*. In the book, Butler also presented a strong objection to 'protoplasm as the only living substance'[81] as suggested by Huxley. Instead, Butler explained that protoplasm could only be accepted as corollary to his memory theory, in contrast to the use of protoplasm as a justification of 'the mindless theory of natural selection'.[82] Butler was very firm on this point. In his opinion, it was not possible to talk about heredity and protoplasm without Lamarckian design. He declared: 'I have said enough to show that in the decade, roughly, between 1870 and 1880 the set of opinion among our leading biologists was strongly against mind.'[83]

Unfortunately, the author of *Life and Habit* was not able to see his theory recognized by the scientific community during his lifetime. He remained an outsider or, citing again Romanes' review, he remained, at least to his contemporary English men of science, 'in no way adapted, either by mental status or mental equipment' to take part in the evolutionary debate.[84]

Notes to Chapter 1

1. Butler, *The Notebooks*, p. 58.
2. See Robert Richards, *Darwin and the Emergence of Evolutionary Theories of Mind and Behavior* (Chicago: University of Chicago Press, 1989), his *The Meaning of Evolution: The Morphological Construction and Ideological Reconstruction of Darwin's Theory* (Chicago: University of Chicago Press, 1993), and his *The Romantic Conception of Life: Science and Philosophy in the Age of Goethe* (Chicago: University of Chicago Press, 2002).
3. George J. Romanes, *Animal Intelligence* (New York: Appenton & Co, 1884), p. 10.
4. The tension between Bergson and Spencer has been explored recently by Hervé Barreau. He discusses, in particular, how Bergson critically used Spencer's work to develop his own theory of evolution. See Hervé Barreau, 'Bergson face à Spencer: Vers un nouveau positivisme', *Archives de philosophie*, 71.2 (2008), 219–43.
5. Butler, *The Notebooks*, p. 57.
6. Charles Darwin, *The Autobiography of Charles Darwin, 1809–1882: With the Original Omissions Restored*, ed. and with an appendix and notes by Nora Barlow (London: Collins, 1958), p. 170.
7. See *The Autobiography of Charles Darwin, 1809–1882*, pp. 167–221; Henry Festing Jones, *Charles Darwin and Samuel Butler: A Step towards Reconciliation* (London: Fifield, 1911); James Paradis, 'The Butler-Darwin Biographical Controversy in the Victorian Periodical Press', in *Science Serialized: Representations of the Sciences in Nineteenth-Century Periodicals*, ed. by Geoffrey Cantor and Sally Shuttleworth (Cambridge: MIT Press, 2004), pp. 307–31.
8. See William Irvine, *Apes, Angels, and Victorians: The Story of Darwin, Huxley, and Evolution* (New York: McGraw-Hill Book Company, 1955), pp. 220–24; Philip J. Pauly, 'Samuel Butler and his Darwinian Critics', *Victorian Studies*, 25.2 (1982), 161–80 (p. 161).
9. Janet Browne, *Charles Darwin: The Power of Place* (London: Cape, 2002), p. 275.
10. Charles Darwin, *The Variation of Animals and Plants under Domestication* (London: Murray, 1868), p. 374.
11. On the place and significance of Newlands' work in the history of the periodic table see W. A. Smeaton, 'Centenary of the Law of Octaves', *Journal of the Royal Institute of Chemistry*, 88 (1964), 271–74 (p. 271); J. W. van Spronsen, 'One Hundred Years of the "Law of Octaves"', *Chymia*, 11 (1966), 125–37.
12. Samuel Butler, *Luck, or Cunning, as the Main Means of Organic Modification? An Attempt to Throw Additional Light upon Charles Darwin's Theory of Natural Selection* (London: Fifield, 1910), p. 13.
13. See Sally Shuttleworth, 'Evolutionary Psychology and *The Way of all Flesh*', in *Samuel Butler, Victorian against the Grain: A Critical Overview*, ed. by James Paradis (Toronto: University of Toronto Press, 2007), pp. 143–69.
14. Shuttleworth, 'Evolutionary Psychology', pp. 147–48.
15. Shuttleworth, 'Evolutionary Psychology', pp. 151–55.
16. See *Samuel Butler, Victorian against the Grain*, ed. by Paradis; Donald R. Forsdyke, 'Heredity as Transmission of Information: Butlerian Intelligent Design', *Centaurus*, 48 (2006), 133–48; 'Samuel Butler and Human Long Term Memory: Is the Cupboard Bare?', *Journal of Theoretical Biology*, 258 (2009), 156–64.
17. Bernard Lightman, 'A Conspiracy of One: Butler, Natural Theology, and Victorian Popularization', in *Samuel Butler, Victorian against the Grain*, ed. by Paradis, pp. 113–43.
18. Lightman, 'A Conspiracy of One', pp. 118–20.
19. Lightman, 'A Conspiracy of One', p. 138.
20. Otis, *Organic Memory*, p. 3.
21. Daniel L. Schacter, *Forgotten Ideas, Neglected Pioneers: Richard Semon and the Story of Memory* (Philadelphia: Psychology Press, 2001), pp. 110–12.
22. Beer, *Darwin's Plots*, p. 5.
23. Butler's opinion concerning professionalism and professional science will be discussed in detail in chapter four.
24. Bernard Lightman, *Victorian Popularizers of Science: Designing Nature for New Audiences* (Chicago: Chicago University Press, 2007), p. 13.

25. Paul White, *Thomas Huxley: Making the 'Man of Science'* (Cambridge: Cambridge University Press, 2003), pp. 51–58.
26. White, *Thomas Huxley*, p. 57.
27. White, *Thomas Huxley*, p. 71.
28. Lightman, *Victorian Popularizers of Science*, pp. 35–37.
29. Samuel Butler, *Life and Habit* (London: Cape, 1910), p. 35.
30. 'Hartmann's *Philosophy of the Unconscious*', review, in *The Modern Review: A Quarterly Magazine*, 5 (1884), 776–81.
31. 'Von Hartmann's Philosophy of the Unconscious', review, in *The Spectator*, 23 (1884), 1111.
32. James Sully, 'The Philosophy of Pessimism', *Westminster Review*, 207 (1876), 59–78 (p. 69).
33. Butler, *The Notebooks*, p. 66.
34. Eva Jablonka, and Mariot J. Lamb, *Epigenetic Inheritance and Evolution: The Lamarckian Dimension* (Oxford: Oxford University Press, 1999), p. 3.
35. Jean-Baptiste Lamarck, *Philosophical Zoology* (London: MacMillan and Co., 1914), p. 113.
36. Ludmilla Jordanova, *Lamarck* (Oxford: Oxford University Press, 1984), p. 76.
37. Pietro Corsi, *The Age of Lamarck: Evolutionary Theories in France, 1790–1830* (Berkeley: University of California Press, 1998), p. 189.
38. See Giulio Barsanti, *Una lunga pazienza cieca: Storia dell'evoluzionismo* (Torino: Einaudi, 2005), pp. 322–27; and Otis, *Organic Memory*, pp. 1–18.
39. See C. Baumann, 'Ewald Hering's Opponent Colors: History of an Idea', *Der Ophthalmologe: Zeitschrift der Deutschen Ophthalmologischen Gesellschaft*, 89.3 (1992), 249–52; R. S. Turner, *In the Eye's Mind: Vision and the Helmholtz-Hering Controversy* (Princeton: Princeton University Press, 1994); Jan Janko, 'Mach and Hering's Physiology of the Senses', *Clio Medica*, 33 (1995), 89–96.
40. Otis, *Organic Memory*, pp. 20–39.
41. See Ewald Hering, 'On the Theory of Nerve-Activity', *The Monist*, 10.2 (1900), 167–87.
42. See review of Ewald Hering, *On Memory and the Specific Energies of the Nervous System*, in *The Monist*, 6 (1895), 634.
43. Butler, *Unconscious Memory*, pp. 63–86.
44. Butler, *Unconscious Memory*, p. 68.
45. Otis, *Organic Memory*, p. 13.
46. Butler, *Unconscious Memory*, pp. 85–86.
47. Barsanti, *Una lunga pazienza cieca*, pp. 306–07.
48. See Jacqueline Carroy and others, 'Les entreprises intellectuelles de Théodule Ribot', *Revue philosophique de la France et de l'étranger*, 4 (2016), 451–64.
49. Vincent Gullin, 'Théodule Ribot's Ambiguous Positivism: Philosophical and Epistemological Strategies in the Founding of French Scientific Psychology', *Journal of the History of the Behavioral Sciences*, 40 (2004), 165–81.
50. Laura Otis and S. Nicolas, *Théodule Ribot: Philosophe breton, fondateur de la psychologie française*, (Paris: L'Harmattan, 2005); Serge Nicolas and Agnès Charvillant, 'Introducing Psychology as an Academic Discipline in France: Théodule Ribot and the Collège de France (1888–1901)', *Journal of the History of the Behavioral Sciences*, 37 (2001), 143–64.
51. Otis, *Organic Memory*, p. 15.
52. Théodule Ribot, *Les Maladies de la mémoire* (Paris: L'Harmattan, 1881), p. 1.
53. Otis, *Organic Memory*, pp. 14–18.
54. Ribot, *Les Maladies de la Mémoire*, p. 6.
55. Ribot, *Les Maladies de la Mémoire*, p. 16.
56. Butler, *Life and Habit*, p. 35.
57. Samuel Butler, *Selections from Previous Works and Remarks on Romanes' Mental Evolution in Animals* (London: Trübner & Co., 1884), pp. 228–29.
58. Samuel Butler, *Essays on Life, Art and Science* (London: Fifield, 1908), p. 240.
59. Butler, *Essays on Life, Art and Science*, p. 240
60. George Romanes, 'Mr. Butler's *Unconscious Memory*', review, in *Nature*, 23.587 (1881), 286–87.
61. Butler, *Remarks on Romanes' Mental Evolution in Animals*, p. 236
62. Butler, *The Notebooks*, p. 53.

63. Butler, *The Notebooks*, p. 59.
64. Butler, *The Notebooks*, pp. 42–46.
65. Roger Robinson, 'From Canterbury Settlement to Erewhon: Butler and Antipodean Counterpoint', in *Samuel Butler, Victorian Against the Grain: A Critical Overview*, ed. by Paradis, pp. 21–44.
66. Butler, *The Notebooks*, p. 66.
67. Butler, *Life and Habit*, p. 33.
68. Samuel Butler, *Luck, or Cunning, as the Main Means of Organic Modification?* (London: Fifield, 1910), p. 9.
69. Butler, *Life and Habit*, p. 198.
70. Butler, *Life and Habit*, p. 198.
71. Butler, *Luck, or Cunning?*, p. 23.
72. Butler, *Unconscious Memory*, p. 53.
73. Butler, *Unconscious Memory*, pp. 55–57.
74. Butler, *Unconscious Memory*, p. 279.
75. Thomas Huxley, *On the Physical Basis of Life* (New Haven, Conn.: The College Courant, 1869), pp. 7–24.
76. George J. Allman, 'Protoplasm and Life', *Popular Science Monthly*, 15 (1879), 721–22.
77. Butler, *Luck, or Cunning?*, p. 125.
78. E. Newton Harvey, 'Some Physical Properties of Protoplasm', *Journal of Applied Physics*, 9.2 (1938), 68–80 (p. 68).
79. Butler, *Unconscious Memory*, p. 62.
80. Butler, *Unconscious Memory*, p. 53.
81. Butler, *Luck, or Cunning?*, p. 127.
82. Butler, *Luck, or Cunning?*, p. 142.
83. Butler, *Luck, or Cunning?*, p. 142.
84. Romanes, 'Mr. Butler's *Unconscious Memory*', p. 286.

CHAPTER 2

Evolution:
From Literature to Science
and Back Again

> They might speak of this by a figure of speech, but they could not see it as a fact. Before this could be intended literally, Evolution must be grasped, and not Evolution as taught in what is now commonly called Darwinism, but the old teleological Darwinism of eighty years ago.
>
> SAMUEL BUTLER, 'God the Known and God the Unknown'[1]

Samuel Butler's writing on science has always been the object of numerous critics and misunderstandings. Since the publication of his first novel *Erewhon*, both reviewers and the general public were unable to fully understand his ideas and writing style. This was mostly due to his engagement with a wide range of different topics. As discussed in the previous chapter, evolution influenced, to a greater or lesser extent, the writing of almost every essay, novel, book, and pamphlet produced by Butler during his lifetime. However, Butler did not always present his scientific ideas in just one form or using one single approach. Instead, as I will outline in this chapter, Butler often mixed the language of science with those of literature, religion, art, and philosophy. His writing style confused his readers, who were unable to understand how Butler's theory of evolution developed organically throughout different publications. It is therefore necessary to re-evaluate Butler's fictional and non-fictional works in the light of his engagement with science, focusing in particular on his science of the mind.

One must execute a close examination of Butler's writing to fully understand his engagement with the scientific and cultural debates of his time. In doing so, I will try to establish how his work fitted (or perhaps tried to fit) within the Victorian evolutionary debate. In particular, I will look at three publications that best represent Butler's work on science, literature, and culture: the novel *Erewhon* (1872), Butler's first book on science *Life and Habit* (1878), and the series of articles 'God the Known and God the Unknown' (1879). In addition, I will present an interpretation of *Erewhon*, based on the claim that part of the novel, 'The Book of the Machines', was not written as a satire. This approach will help us to see how Butler's theory of 'unconscious evolution' was presented to the Victorian public long before the publication of *Life and Habit*. In my reading, the novel is, instead, a piece of popular science designed to express the admiration of the author for Darwinism and the idea of evolution.

Framing Butler's Writings on Science

One of the key aspects that has always influenced any interpretation of Butler's work was his inability to focus on just one particular field. As suggested by the Irish playwright, critic, and political activist George Bernard Shaw (1856–1950), Samuel Butler's work was the product of one of the most varied and eccentric careers of the Victorian period.[2] Butler explored a series of different genres ranging from traveller's guides to foreign countries to utopian novels. He wrote about art, science, and literature while also venturing into the composition of musical scores influenced, especially, by George Frideric Handel (1685–1759). Fortunately, although Butler was not a hoarder, he destroyed precious little. Notebooks, old manuscripts, torn-out pages, annotated scientific books were all stashed away. In the last few years, scholars have embarked on a wide-ranging series of discoveries in which new material has shown the eccentricity of his style of writing. Already considered an eccentric Victorian by his contemporaries, Samuel Butler's work, as presented by Paradis in the introduction of his edited collection, is the product of a polymath who loved to experiment with different ideas and fields.[3]

In order to question and explore how Butler mixed different genres, it is better to start from some preliminary assumptions concerning the importance that genres play in any fictional or non-fictional piece of writing. Critical interpretation of the work of writers, novelists, and poets often involves investigation into literary categories, such as style, topic, plot, content, tone, metaphor, theme, narrative, persona, time, culture, and genre. As indicated by Christina Myers-Shaffer's *The Principles of Literature* (2000), the examination of such categories is of primary importance in understanding a given piece of writing and offers the 'opportunity to look into the hearts and minds of people who lived in a different time and place'.[4] Indeed, in line with what Myers-Shaffer suggests, my aim is not to challenge the plot or arguments presented by Butler in his writing. It is, instead, to analyse how the Victorian scientific and cultural contexts directly influenced Butler's narratives. Therefore, extracts from both his fictional and non-fictional writings will be discussed and examined in relation to the following aspects: 'meaning', 'form', 'tone', and 'language'. All of these aspects are considered as the basis of the interpretation of any piece of writing. However, as observed, once again by Myers-Shaffer, they can only work together for unity to produce the unique blend of a specific piece of writing.[5]

This is where Butler's writings start to be problematic. As many biographical studies have noted, Butler was never completely sure about his vocation. He changed opinion about himself and his profession several times during his life. Peter Raby, in his biography of the writer, pointed out the impossibility of distinguishing any single genre within Butler's writing.[6] Neither a scientist, novelist, artist, nor philosopher, the main peculiarity of this eccentric Victorian was not to be a specialist in any field.[7]

In addition, Butler's inadequate knowledge of any specific scientific discipline, combined with the naive desire to be famous and respected as a professional, made his books difficult to read and categorize. It is no surprise, therefore, to see how Butler's writing on science was misunderstood by both professional and lay readership alike. In 1985, the American literary scholar Philip Cohen explained that the rise and fall of Butler's personal and literary fame was due to his mixing literature with science, which made him a 'philosopher of common sense' rather than a professional scientist or a novelist.[8] Almost twenty years later, James Paradis, in the introduction to *Samuel Butler, Victorian against the Grain*, came to the same conclusion. He maintained that 'if Butler was important to writers', referring to his reception among novelists such as George Bernard Shaw, Virginia Woolf, and F. Scott Fitzgerald, 'his reception among critics was controversial, beginning with — perhaps especially with — his contemporaries, who were often puzzled or offended by his free-thinking, audience-baiting irony.'[9] Before Paradis, Lee Holt's article 'Samuel Butler up to Date' (1960) similarly explained that the critical reception of Butler's work was '[g]enerally blinded to real value by their [Victorian reviewers] moral, religious, scientific and literary preconceptions'.[10] Victorian reviewers were, indeed, unable to look at the intrinsic significance of Butler's ideas, often focusing just on the satirical tone of his writings.

This was due mostly to Butler's publishing politics. Looking at his publications in chronological order, it is even difficult to divide them into precise categories or find a way to link them all together. For example, in the 1880s — while Butler was writing *The Way of All Flesh* (1903), his most controversial and successful novel — he was studying and translating foreign essays on evolution as well as publishing his major works in the field. Furthermore, when he was having his quarrel with Darwin's family, trying to establish himself as an amateur scientist, he was also writing a book entitled *Alps and Sanctuaries of Piedmont and the Canton Ticino* (1881) a travel guide about the culture and religious art of northern Italy.

To explain his desire to work in different fields, Butler left us with a note in his notebooks where he tried, in his own way, to define his peculiar publishing strategy:

> I never make them: they grow; they come to me and insist on being written, and on being such and such. I did not want to write Erewhon, I wanted to go on painting and found it an abominable nuisance being dragged willy-nilly into writing it. So with all my books — the subjects were never of my own choosing; they pressed themselves upon me with more force than I could resist. If I had not liked the subjects I should have kicked, and nothing would have got me to do them at all. As I did like the subjects and the books came and said they were to be written, I grumbled a little and wrote them.[11]

Butler was obviously aware of his nature of being a polymath. Indeed, while writing and conducting research on evolution, he was also painting and studying arts, writing musical pieces, and engaging with controversial literary issues such as the order of Shakespeare's sonnets or whether the *Odyssey* was written by a young Sicilian woman. This clearly impacted on how Butler's work was posthumously defined. There are several examples to cite here which can help us understand how

Butler's peculiar use of different writing styles and engagement with different fields was explained in various ways by scholars throughout the twentieth century.[12] It is, therefore, necessary — before moving onto the analysis of Butler's literary writing on science — to look at how the critical reception of his work changed since the late nineteenth century.[13] This can shed some additional light on the disagreements that his work generated among scholars.

As with other Victorian writers, posthumous interpretations of Butler were, very often, and for various reasons, in contrast with one another. However, they all agreed on one point: Butler's work is difficult to categorize into a single literary or scientific genre. For instance, it is worth looking at one of the first critical studies on Butler, *The Earnest Atheist* (1936), by the journalist Malcolm Muggeridge.[14] Although not the first study published on Butler (by that time Henry Festing Jones (1851–1928) had already published a two-volume biography on his work), Muggeridge's study offers probably the first attempt to analyse Butler's writings critically. In the introduction of his book, Muggeridge made an important point about Butler's work: the English writer was not an anti-Victorian (as his contemporaries considered him); he was instead the ultimate Victorian. An individual tormented by the need to escape from the reality of his own existence.[15]

For Muggeridge, Butler's need to think ahead of his time was the key element of his written output. Indeed, as Muggeridge pointed out, Butler always tried to escape from tradition, criticizing and taking to the extreme not only Victorian cultural values but also his contemporary society's social, political, and scientific conventions. However, and this is important, Muggeridge also explained that Butler cannot be considered as either simply a Victorian writer or an amateur scientist. The Victorian writer was, instead, a thinker working somewhere in between literature and science. If we look at what Muggeridge writes about Butler, we immediately notice how different his opinion was compared to previous critical accounts:

> Samuel Butler must be regarded as one of the most significant figures of the latter part of the last century. His own generation ignored him. His fame was almost wholly posthumous. In so far as he was known at all in the flesh he was as an oddity, an eccentric with a number of queers in his bonnet, as that the Odyssey was written by a woman, that the credit for formulating the theory of evolution must go rather to Erasmus Darwin, Buffon and Lamarck than to Charles Darwin, and that habit, not chance, was the chief factor in producing variations.[16]

The quotation above reveals an important point about Butler's approach to writing. Butler lived on the edge of eccentricity and oddity, and his work was criticized and ostracized by both scientists and non-scientists because of his attitude. However, a 'few years later his reputation had swollen to immense proportions'.[17] Muggeridge obviously referred to the fame that Butler's literary and scientific ideas were acquiring in the early twentieth century. For now, it is important to recognize that in the 1920s, Butler's work started to be recognized and discussed all across Europe.

There is another interesting example worth looking at, namely a partial account of Butler's scientific work which comes from an old-school historian of science: William Irvine. In 1955, Irvine published the now classic *Apes, Angels and Victorianism*. The volume, which aimed at being a biographical account of the lives and works of Charles Darwin and Thomas Huxley, provided an overview of the historical context, networks, and family gossip of the main protagonists of the Victorian evolutionary debate. Although references to Butler in the volume are rather limited, Irvine's book still provides an interesting account of his place in the historical debate of the 1950s. Interestingly, Butler's work was considered, at best, as no more than an anecdotal episode in the historical examination of Darwin's work. Indeed, Irvine described the author of *Erewhon* as:

> He [Butler] was suspicious, critical, alert, witty, devious, elusive. He also had the satirist's power of turning his victims into vivid, plausible villains who thoroughly deserved the savagery of his pen. The Darwin of *Luck or Cunning?*, like the Theobald Pontifex of *The Way of All Flesh*, causes the reader to burn with moral indignation.[18]

Thus, in Irvine's portrayal of Victorian London, Butler was just reduced to the perfect anti-Victorian. Of course, Irvine's interpretation of Butler does not reveal much about the latter's scientific ideas or relationship with Darwin. For Irvine, Butler did not have a prominent role in the evolutionary debate, and his writing on science was at best an unsophisticated and amateur attempt to talk about evolution. Therefore a proper examination of his writings on science was, for Irvine, simply not necessary. Indeed, by focusing on just the quarrel between Darwin and Butler, Irvine totally dismissed Butler's own scientific ideas, explaining how the disagreement was just the result of Butler's personal attack on the father of evolution. Irvine writes:

> By a very romantic logic of his own, Butler had found Darwin to be first superhuman, then human, and finally inhuman. That inhumanity needed only to be demonstrated by an overt act. Of course it was, almost at once.[19]

In this respect, Irvine's work represents just another example of how Butler was considered as an outsider due to his personal attitude and inability to professionally contribute to the evolutionary debate.[20] Other studies published around the middle of the twentieth century, including P. N. Furbank's *Samuel Butler, 1835–1902* (1948), Stanley Bates Harkness' *The Career of Samuel Butler, 1835–1902: A Bibliography* (1955), and Holt's *Samuel Butler* (1963), provide a similar biographical account of the life and work of the Victorian writer, without adding anything new to our understanding of his writing on science. This is not at all surprising, as at the time Butler's contribution to the scientific debate had been long forgotten and ignored.

To see a change in this critical understanding of Butler's ideas, we have to wait for the more recent works of Elinor Shaffer and James Paradis. Both Shaffer's *Erewhons of the Eye* (1988) and the edited collection *Samuel Butler, Victorian against the Grain* (2007) by Paradis, explore aspects of Butler's life and writings traditionally ignored by classic scholarship. This change is clear from the introduction of Shaffer's book that states: 'all accounts so far given of Samuel Butler are truncated: they are savage amputations of the limbs of his life.'[21] What Shaffer offered is therefore an opportunity to look into some aspects of Butler's work which, to date, had been overlooked by previous scholars. Indeed, Shaffer's work aimed to reintegrate

Butler's various writings 'into a full account of multifarious activity as writer, painter, and author of pungent and far-reaching critiques of literature, art, religion and science'.[22]

This need to look at the work of Butler as a whole was also at the heart of the collection of essays edited by Paradis in 2007. Indeed, in introducing the volume, Paradis explains:

> Samuel Butler (1835–1902), Victorian satirist, critic, and visual artist, possessed one of the most original and inquiring imaginations of his age. [...] Evolutionary free-thinker, he rejected natural selection and traditional natural theology alike in a series of evolutionary studies [...] that placed evolutionary thinking within a new historical framework and that assessed, in a neo-Lamarckian context, the role of memory in shaping the organism.[23]

Starting from the necessity of rediscovering some aspects of Butler's research and his position within the Victorian debate, the collection of essays provides a new portrayal of the different sides of Butler's persona and how his being controversial and against tradition was reflected in his writing production. Indeed, for Paradis, Butler had to be considered as 'Victorian England's ultimate polymath, an artistic and intellectual ventriloquist who assumed an extraordinary range of roles — as satirist, novelist, evolutionist, natural theologian, travel writer, art historian, biographer, classicist, painter, and photographer'.[24]

Science, Literature, and Evolution: *Erewhon*, *Life and Habit*, and 'God the Known and God the Unknown'

In order to understand the impact that working as a polymath had on the development of Butler's work and ideas, it is necessary to start from the role played by genres in his fictional and non-fictional writings. As Beer clearly explains in *Open Fields* (1996), the study of genres involves conflict and transformation as much as mutual understanding and reconciliation.[25] Indeed, as James Secord also discusses, the study of genres is part of the study of the interaction between different cultures, the history of books and production of knowledge and their development within a specific readership.[26]

The interaction between different genres and, more generally, between science and literature is particularly important in framing Butler's writing on science. As is well known, the second part of the nineteenth century was a period of experiments and revolutions for science. The life and work of scientists and explorers became fashionable among the general public. Certainly Darwin's work, which encompassed collecting specimens from all around the globe, demonstrated how one single theory could produce both a strong scientific statement and an exciting read for the general public. Darwin's hypothesis of 'natural selection' helped to cross social and international boundaries by showing how the language of evolution was not only changing science but also influencing society as a whole.

As explored by Gillian Beer in *Darwin's Plots*, Victorian culture after the publication of *Origin of Species* not only started to reflect and use ideas relating to evolution but also Darwin's scientific language and desire to explore unknown territories.[27] Indeed, in the nineteenth century, Beer explains, scientific language and narratives were moving 'rapidly and freely to and from between scientists and non-scientists' and this was not only a one-way process.[28] On the one hand, well-known scientists such as Charles Darwin, Claude Bernard, and Charles Lyell often employed metaphors and analogies from the literary world in order to explain complex scientific ideas to their audiences. On the other hand, novelists heavily relied on scientific discoveries, ideas, and terminology in crafting their stories. Indeed, from the 1860s, the use of scientific terms in literature produced a revolution in the cultural vocabulary and contributed to increasing the scientific literacy of the general public. Words such as 'evolution', 'degeneration', 'observation', 'research', 'natural selection', and 'exploration' started to be used actively by novelists.

Similarly, as James Secord explains in *Victorian Sensation*, the incredible success and reception of Robert Chambers' *Vestiges of Natural History of Creation* was due to a variety of factors including the growing interest in natural science in the cultural and philosophical debate of the mid-nineteenth century. Secord uses the public as a category to explore the reception of a scientific book in the period and explains that:

> placing reading at the centre of a history opens up general possibilities for understanding what happens when we read [...]. It unites an interpretation of words on the page with an understanding of the physical appearance and genre of work in which it is marked and discussed.[29]

Placing Butler's writing style and his use of different genres at the core of our examination of his engagement with science can shed more light on the development of his theory of evolution. One of the major challenges with Butler's written works has always been the difficulty of understanding the relationship between content, style, and format in his books. Let us take as an example his engagement with Darwin's work on evolution. From the beginning of his writing career to his last work on science published in 1890, Butler placed the discussion of Darwinian evolution at the centre of a debate which also encompassed the broad study of art, music, philosophy, history, and theology. Butler, as many other polymaths of his generation, was aware of how the use of different genres and styles of writing directly influenced the reception of an author among different readerships. In 1893, Butler wrote about the Victorian readership:

> People between the ages of twenty and thirty read a good deal, after thirty their reading drops off and by forty is confined to each person's special subject, newspapers and magazines; so that the most important part of one's audience, and that which should be mainly written for, consists of specialists and people between twenty and thirty.[30]

At the time Butler was writing his main fictional and scientific works, he was conscious of the role played by general and specialist audiences in making a publication successful. As indicated above, Butler also paid particular attention to target anyone with a special interest in science or between the age of twenty and thirty. Therefore, it is of no surprise to see how in all of his books, Butler always tried to speak with both the professional men of science and the less educated men of the street.

Butler's first novel *Erewhon* was published anonymously in 1872. His decision to remain unnamed as author was due to two main reasons. First, this practice was a typical strategy adopted by Victorian editors in publishing the work of new and unknown writers. Second, Butler's decision to publish his work anonymously was due to a sense of insecurity about the possible reaction of his family to the topic of the novel.[31] Since his childhood, Butler had had a very complex relationship with his father due to differing ideas concerning his future career. As per family tradition, after completing his studies, Butler was expected to join the church; instead, he decided for a different future by moving to New Zealand to become a farmer. Needless to say that this decision to move to another country created a tension between Samuel and his father. An account of Butler's childhood and complex relationship with his family is narrated in the semi-biographical novel *The Way of all Flesh*. The desire of the author to not share the content of *Erewhon* with his father is clearly stated in a letter written by Samuel's father Thomas on 29 May 1872:

> Dear Sam,
> I shall take your advice and not read your book. It would probably pain me and not benefit you. I do not the least object to your putting your name to it tho' I may not value the éclat. The grief is that our views should be so wide asunder. Perhaps the book might pain me less than your letter leads me to infer. I gladly give it the benefit of the doubt.
> Your affectionate father,
> T. Butler[32]

The first edition of *Erewhon* was very successful, but from the second edition onward sales fell sharply (by around 90%) once Butler's name was announced. This was because Butler was not famous enough to be recognized and appreciated by Victorian readers. By 1893, the novel sold only 3864 copies and it was the most popular book of Butler's career.[33]

The reason behind *Erewhon*'s initial success was due to a simple misunderstanding: at first, Butler's book was considered by Victorian readers to be the sequel to Edward Bulwer Lytton's 1871 novel *The Coming Race*. The two novels presented some similarities. For example, characters were given similar names: the heroine of *The Coming Race* was called Zee; in *Erewhon*, the main female character was given the name of Zelora, and in the second edition Zelura. Furthermore, the content and style of writing, especially in relation to the use of satire applied to Victorian culture, seemed to be the product of the same author. In *Erewhon*, Butler critically engaged with some of the main topics being discussed in 1870s London: evolution in addition to a mixture of religious, social, philosophical, and anthropological ideas. The novel, set in a utopian New Zealand countryside, employed satire to describe a society of 'savages' to give, as many other Victorian utopian novels did, an 'ironic description of English assumptions' about non-European populations.[34] Nevertheless, *Erewhon* also aimed to provide a dramatic prophecy about the future of Victorian society.

As noted by Sue Zemka, *Erewhon* is and has always been classified as a utopian novel although it presents some divergences from the classic utopian literature by favouring a more anthropological type of satire.[35] However, certain sections of Butler's 'Book of the Machines' can — and should be — read as a non-satirical philosophical essay. This is a highly critical and controversial aspect of the novel which also raises a key epistemological question: is it possible to read *Erewhon* as an example of Victorian satire and utopian writing and, at the same time, a speculative investigation of organic and inorganic evolution? Answering this question is crucial for providing a new understanding of the novel. It enables us to establish why Butler chose to write a novel which attempted to represent evolution in a humorous and creative manner.

Similarly, it is also worth questioning why Butler decided to develop and incorporate his earlier short essays 'Darwin among the Machines' and 'Lucubratio Ebria' into a fictional work as opposed to turning them directly into a non-fiction book in the manner of *Life and Habit*. To answer this, we need to look more closely at the scientific and philosophical allegories used in Butler's writing. My claim is that Butler actively used the novel as an experiment to test his own theory of evolution.

In the essay 'Butler, Memory and the Future', Gillian Beer argues that the desire to provide a prophecy was one of the central points of *Erewhon*, especially within the 'Book of the Machines'.[36] Beer wrote:

> Indeed, futurity has borne out some of Butler's most important arguments: the 'Book of Machines' in *Erewhon* not only foresees the development of computers but prognosticates their miniaturizations, their capacity to undertake computations that outrun human capacities, and their capacity for self-reproduction.[37]

In developing the idea of the machinery's evolution, Beer suggests, Butler applied the real essence of evolution to machines as an example of inorganic life.[38] Beer's interpretation has merit, but it misses the central point of the 'Book of the Machines': whilst Butler's writing provides a prophecy of the future of the human race, it also presents some interesting considerations about Darwin's idea of natural selection.

It follows, that in order to fully understand the philosophical nature of the novel, a close analysis of Butler's technological analogies becomes necessary. Indeed, in the 'Book of the Machines', the content and style of writing shifts between that of a fiction (in a satirical and prophetical genre) and non-fiction essay. For instance, Butler philosophically challenged the notion of 'conscious' and 'unconscious evolution' posing questions such as: 'Where does consciousness begin, and where does it end? Who can draw the line? Who can draw any line? Is not everything interwoven with everything? Is not machinery linked with animal life in an infinite variety of ways?'[39] However, in the same paragraph, Butler, in order to reinforce the credibility of his argument and to appeal to a more sophisticated readership, decided to employ a specific scientific language. An example of his use of pseudoscientific terminology can be found in the first part of 'The Book of the Machines' where Butler writes:

> no animal has the power of originating mechanical energy, but that all the work done in its life by any animal, and all the heat that has been emitted from it, and the heat which would be obtained by burning the combustible matter which has been lost from its body during life, and by burning its body after death, make up altogether an exact equivalent to the heat which would be obtained by burning as much food as it has used during its life, and an amount of fuel which would generate as much heat as its body if burned immediately after death.[40]

The extract above presents the style and terminology of a scientific essay. Indeed, it is nothing less than a quotation, although with some small differences, not properly acknowledged, of a paper by William Thomson (1824–1907) entitled 'Origin and Transformation of Motive Power':

> It appears certain from the most careful physiological researches, that a living animal has not the power of originating mechanical energy; and that all the work done by a living animal in the course of its life, and all the heat that has been emitted from it, together with the heat that would be obtained by burning the combustible matter which has been lost from its body during its life, and by burning its body after death, make up together an exact equivalent to the heat that would be obtained by burning as much food as it has used its life, and an amount of fuel that would generate as much heat as its body if burned immediately after birth.[41]

The plagiarism of Thomson's essay offers us an insight into Butler's attempt to popularize evolution in the novel. Indeed, Butler's aim was far from the simple desire of writing a novel or becoming a professional storyteller. Instead, he attempted to make his work more scientifically credible. The plagiarism of Thompson further reveals his distinctive approach to both literature and science. For Butler, evolution had to be treated as a serious philosophical, literary as well as scientific topic.

In Butler's novel, it is difficult to clearly distinguish between his fiction and non-fiction writing styles. By examining the use and significance of the notion of 'evolution' in the novel, we can see how it could be deployed to express different meanings. Firstly, Butler's description of the population of *Erewhon* uses satire to show how concepts like 'evolution', 'extinction', and 'race', normally used within the natural context, could be applied to a human society, where living and evolving 'machines' have become dangerous instruments in the hands of humankind. It is no surprise then to see that Butler described humankind as a 'machinate mammal' using 'machines as a supplementary limb'.[42] This anthropological interpretation of human evolution employs a fictional analogy to describe and criticize how humankind is sometimes too dependent on its own technology. This claim is supported by several examples in which the Victorian writer tried to show how machines modify human life by replicating artificial limbs:

> Man has now many extra-corporeal members, which are of more importance to him than a good deal of his hair, or at any rate than his whiskers. His memory goes in his pocket-book. He becomes more and more complex as he grows older; he will then be seen with see-engines, or perhaps with artificial teeth and hair: if he be a really well-developed specimen of his race, he will be furnished with a large box upon wheels, two horses, and a coachman.[43]

The quotation above must be read as a complement to Beer's interpretation of the novel as a prophetical examination of a possible future of human society. Indeed, by examining the relationship between evolution and machines, it becomes easier to detect the *escamotage* Butler employed to question the cultural implications of Darwin's hypothesis of natural selection when applied to the mechanical kingdom. This argument has been explored in Joshua A. Gooch's article 'Figures of Nineteenth-Century Biopower in Samuel Butler's Erewhon', where the novel is read in the fashion of social Darwinism as applied to the notion of biopower. Gooch's article considers Butler's use of evolution in the novel as an expression of 'life's management at the level of population'.[44] Gooch argues that certain aspects of Michel Foucault's idea of power were anticipated by Butler in *Erewhon*. However, *Erewhon* did not challenge evolution in its political and economic contexts. This is why Butler justified his hypothesis of mechanical evolution by creating a fictitious 'man of science'. In the novel, Butler insisted upon the following point:

> I do not know how he has found this out, but he is a man of science — how then can it be objected against the future vitality of the machines that they are, in their present infancy, at the beck and call of beings who are themselves incapable of originating mechanical energy?.[45]

The use of a 'man of science' allowed Butler to justify the idea that machines, like animals, can derive energy from food by adopting and applying the laws of thermodynamics to the metaphor of mechanical life. Butler, of course, took it to an extreme:

> The main point, however, to be observed as affording cause for alarm is, that whereas animals were formerly the only stomachs of the machines, there are now many which have stomachs of their own, and consume their food themselves. This is a great step towards their becoming, if not animate, yet something so near akin to it, as not to differ more widely from our own life than animals do from vegetables.[46]

By taking a closer look at the quotation above, there is again reason to suggest that Butler's writing can be read on two different levels. On the one hand, there is the fictional element to the text where animals are depicted as metaphorical stomachs of machines. On the other hand, this anthropomorphization of machines can also be read as a protagonist in a philosophical debate aimed at stressing the limits and potential of mechanical evolution.

The metaphor of 'mechanical evolution' was first used by Butler in 1863 in 'Darwin among the Machines', where evolution was discussed *via* a philosophical examination of the metaphor of 'artificial life'. The article suggests the likelihood of a future war between humans and machines for establishing 'what sort of creature man's next successor in the supremacy of the earth is likely to be'.[47] Again Butler took this argument to its philosophical extreme by suggesting:

> Day by day, however, the machines are gaining ground upon us; day by day we are becoming more subservient to them; more men are daily bound down as slaves to tend them, more men are daily devoting the energies of their whole lives to the development of mechanical life.[48]

'Darwin among the Machines' was clearly written as a clever 'joke' but introduced an important question regarding the hereditary process and discussed, for the first time, the role of memory as part of Butler's idea of evolution.

In 1865, with 'Lucubratio Ebria', Butler's metaphor of 'evolving machines' morphed into an anthropological interpretation of human evolution, whereby the evolution of the human body was explained through the use of 'artificial limbs'. In this short article, Butler questioned the complex relationship between our body and the technology we utilize; he arrived at the conclusion that human evolution was different from that of animals because it was mediated by the use of various technological tools. For example, in the article Butler talks about clothes to protect our body from the cold, spectacles to help with reading, and notebooks as a memory box. However, as Butler explains, the relationship between humans and their technologies is far beyond the simple use of tools: 'men are not merely the children of their parents, but they are begotten of the institutions of the state of the mechanical sciences under which they are born and bred.'[49] Butler goes as far as to claim that:

> We are children of the plough, the spade, and the ship; we are children of the extended liberty and knowledge which the printing press has diffused. Our ancestors added these things to their previously existing members; the new limbs were preserved by natural selection, and incorporated into human society; they descended with modifications, and hence proceeds the difference between our ancestors and ourselves.[50]

With regard to the relationship between the 1865 article and 'Darwin among the Machines' Butler wrote:

> It is a mistake, then, to take the view adopted by a previous correspondent of this paper; to consider the machines as identities, to animalise them, and to anticipate their final triumph over mankind. They are to be regarded as the mode of development by which human organism is most especially advancing, and every fresh invention is to be considered as an additional member of the resources of the human body.[51]

The two articles, although dissimilar in terms of content, style, and ambition, were linked by Butler's emphatic use of Lamarck's hypothesis of the 'inheritance of acquired characteristics'. Butler clearly understood that Lamarck's philosophy of evolution was primarily progressive, although it did involve some divergences.[52] According to Lamarck's *Philosophie zoologique*, in nature there is no extinction, and evolution determines the passage from a simple to a complex form of life based on a continuous reproduction of certain acquired characteristics.[53] Lamarck's hypothesis of evolution, as largely discussed in the previous chapter, is based on the inheritance of modifications produced by a guided process of 'use and disuse' of organs. It is therefore of no surprise to see how in 'Lucubratio Ebria' Butler insisted on explaining evolution by employing an internal 'force' that shows:

> Very clearly that each species of the animal and vegetable kingdom has been moulded into its present shape by chances and changes of many millions of years, by chances and changes over which the creature modified had no control whatever, and concerning whose aim it was alike unconscious and indifferent.[54]

The similarity between 'Lucubratio Ebria' and Lamarck's *Philosophie zoologique* is self-evident although Butler did not cite Lamarck directly. At the time, Butler's knowledge of the French naturalist was based on what he was able to read in Darwin's work.

The Lamarckian link between the New Zealand articles and *Erewhon* is not difficult to see either, although these publications were intrinsically different. Both 'Darwin among the Machines' and 'Lucubratio Ebria' were, again, short articles — three pages long and without any clear philosophical ambition. In comparison, the 'Book of the Machines' was a twenty-five-page essay in which Butler tried to construct a complex philosophical argument. In *Erewhon*, Butler attempted to highlight the presence of a Lamarckian 'inheritance of acquired characteristics' in vegetables, animals, and machines by labelling it as 'unconscious evolution'. Butler's 'unconscious memory', like Lamarck's 'inheritance', was founded on a process of unconscious biological reproduction and preservation of information from one generation to the next. In addition, the use of Lamarckism proved to be perfect for bridging the need of a philosophical examination of the notion of unconsciousness with the materialism advanced by Darwinian evolution.

For example, in the novel Butler deploys another scientific figure, in the guise of an old *Erewhonian* professor of botany, to define the preservation of information in vegetables:

> It will become unconscious as soon as the skill that directs it has become perfected. Neither rose-seed, therefore, nor embryo should be expected to show signs of knowing that they know what they know — if they showed such signs the fact of their knowing what they want, and how to get it, might more reasonably be doubted.[55]

A further example can also be used to illustrate the influence of Lamarckism. In *Erewhon*, chapter 27, Butler engaged with a philosophical debate regarding the rights of animals and vegetables by proposing them as equal.[56] Leaving aside the philosophical implications of this chapter, it is also noticeable that Butler employed Lamarckism to present an argument, once again, about heredity. Butler explained that since animals and plants both have a common ancestry and evolved through the same hereditary process, they should have the same right to life. The reference to Lamarck is enclosed in Butler's use of 'unconscious memory', settled in any living entity 'beyond further power of question'.[57] Indeed, for Butler, memory has to be seen as the key component of the hereditary process which gives to any organism the right 'of knowing what they do, provided they do it, and do it repeatedly and well, the greater proof they give that in reality they know how to do it and have done it already on an infinite number of past occasions'.[58] This Lamarckian ability of 'knowing' how to progress from one generation to the next, and the prospect of being able to trace a common ancestor, put plants and animals on an equal footing for the inhabitants of this utopian land.

As indicated earlier, Butler's engagement with evolution in the novel makes the identification of a single genre for *Erewhon* problematic. Indeed, the 'Book of the Machines' could be read simply as a satirical interpretation of Darwinian evolution. This confusion is confirmed by the general agreement of reviewers who defined it as an example of Victorian satire. Butler was, of course, aware of this misunderstanding, and in the second edition of the novel he responded to *Erewhon*'s negative reviewers by stating:

> I regret that reviewers have in some cases been inclined to treat the chapters on Machines as an attempt to reduce Mr. Darwin's theory to an absurdity. Nothing could be further from my intention, and few things would be more distasteful to me than any attempt to laugh at Mr. Darwin; but I must own that I have myself to thank for the misconception, for I felt sure that my intention would be missed, but preferred not to weaken the chapters by explanation, and knew very well that Mr. Darwin's theory would take no harm.[59]

In *Erewhon*, Butler did not make any jokes at the expense of the concept of 'natural selection'. Instead, he used a satirical style of writing in demonstrating the real and intrinsic force of Darwin's argument. In a letter to Darwin dated 11 May 1872, Butler explained the real idea behind the 'Book of the Machines' by saying: 'I am sincerely sorry that some of the critics should have thought I was laughing at your theory, a thing which I never meant to do and should be shocked at having done.'[60]

Another example which can shed some additional light on Butler's complex mix of different writing styles and topics is his first book on science: *Life and Habit*. In contrast to *Erewhon*, *Life and Habit* explored the idea of 'unconscious memory' in a more scientific form. Butler's intention to write a scientific book is confirmed by a note published in the *New Quarterly Review*. There, Butler, talking about his career and written work, defined *Life and Habit* as an *'opus prima'* and his first attempt to explain the correlation between memory and heredity in evolution.[61] However, despite being considered by Butler his first work specifically written on evolution, *Life and Habit* presents a clear continuity with his previous writings.

Proof of this continuity can be found, once again, in Butler's notebooks. In the section entitled 'The Germs of Erewhon and of Life and Habit', the writer explains how the genealogy of his idea went through a long process, which started when he was living in New Zealand. For Butler, the starting point of both *Erewhon* and *Life and Habit* was his admiration for Darwin's *Origin of Species*.[62] When Darwin's work was published, Butler became one of the many people fascinated by his theory of natural selection. On this particular aspect of Butler's biography, Henry Festing Jones wrote:

> They are interesting as showing that Butler was among the earliest to study closely the *Origin of Species*, and also as showing the state of his mind before he began to think for himself, before he wrote *Darwin among the Machines* from which so much followed; but they can hardly be properly considered as germs of *Erewhon* and *Life and Habit*. They rather show the preparation of the soil in which those germs sprouted and grew; and, remembering his last remark on the subject that 'it was all very young and silly,' I decided to omit them. The *Dialogue* is no longer lost, and the numbers of the *Press* containing it and the correspondence that ensued can be seen in the British Museum.[63]

As I will discuss in the next chapter, 'The Dialogue' was very well received by the local community and even got the attention of Darwin himself. Returning to the similarity between *Erewhon* and *Life and Habit*, in the 1901 revised edition of the novel, Butler explained that:

> The first part of Erewhon written was an article headed 'Darwin among the Machines' and signed 'Cellarius.' It was written in the Upper Rangitata district of Canterbury Province (as it then was) of New Zealand and appeared at Christchurch in the *Press* newspaper, June 13, 1863. A copy of this article is indexed under my books in the British Museum catalogue.[64]

Additionally, specifically about the common origin of *Erewhon* and *Life and Habit*, Butler pointed out that:

> A second article on the same subject as the one just referred to appeared in the *Press* shortly after the first, but I have no copy. It treated machines from a different point of view and was the basis of pp. 270–274 of the present edition of *Erewhon*. This view ultimately led me to the theory I put forward in *Life and Habit*, published in November, 1877. I have put a bare outline of this theory (which I believe to be quite sound) into the mouth of an Erewhonian professor in Chapter XXVII of this book.[65]

As indicated in the quotation above, continuity of content between these two volumes went far beyond this simple New Zealand common root. However, the Victorian writer was also unsure about the best way to present his theory of evolution to the general public. Butler wrote in a letter (18 February 1876) to his father: 'My present literary business is a little essay some 25 or 30 pp. long, which is still all in the rough and I don't know how it will shape.'[66] Indeed, in 1876, *Life and Habit* was nothing more than a long paper about memory, identity, heredity, and reproduction, and Butler was not sure about the shape it would eventually take.[67]

There is an additional point to make about the continuity between Butler's novel and first book on science. If in *Erewhon* Darwinian evolution was explored and challenged in a fictional way, in *Life and Habit* Butler's aim was almost the opposite. As discussed above, the main problem for interpreting the significance, content, and tone of *Erewhon* was best represented by 'The Book of the Machines' and it being a non-fictional essay in a novel. However, despite the different type of publication and aim, *Life and Habit* generated similar problems among the Victorian readership, who were unable to see the book as a serious attempt to contribute to the vibrant nineteenth-century evolutionary debate. They questioned if *Life and Habit* could be considered as a scientific book or rather another satirical publication from the author of *Erewhon*.

According to Butler, *Life and Habit* was written with the aim of seriously discussing evolution, especially in relation to some of the ideas he already introduced in *Erewhon*. However, *Life and Habit* was also written with the:

> smallest pretension to scientific value, originality, or even to accuracy of more than a very rough and ready kind — for unless a matter be true enough to stand a good deal of misrepresentation, its truth is not of a very robust order, and the blame will rather lie with its own delicacy if it be crushed, than with the carelessness of the crusher.[68]

There is, therefore, a clear contrast between the scientific aim of the book and

Butler's personal statement about its value. In writing *Life and Habit*, Butler mixed together two separate vocabularies and styles of writing. Butler used scientific terms such as 'evolution', 'instinct', 'processes of digestion', 'the action of the heart', and the 'oxygenation of the blood', alongside Latin expressions such as '*de novo*', '*cogito ergo sum*', and '*summum bonum*'. In addition, references to classical art including descriptions of the *Venus de Milo*, the *Discobolus*, and the *Saint George* of Donatello were discussed next to quotations from Carpenter's *Mental Physiology*, Darwin's *Origin of Species,* and other key scientific works from the period. This mixture of writing styles clearly confused the Victorian general readership as well as the professional one.

The reception of Butler's first book on science was disastrous. Not only did the scientific community reject the idea of memory as a form of heredity, but also Butler himself was ridiculed. He was defamed as a writer without any scientific experience who tried to work as a professional without having the competence of being one. There are, for instance, two reviews of *Life and Habit*, which can shed some light on the general misunderstanding generated by Butler's writing.

On the one hand, the *Saturday Review* (26 December 1878) dismissed any scientific value to *Life and Habit* by saying:

> The author of *Erewhon* might have been expected to write a fanciful book, and he had done so; but he has also shown himself capable of more than mere fancies. The reviewer disclaimed, indeed, any seriously scientific initiation, as well as any special scientific knowledge.[69]

On the other hand, another review, this time from the *Daily News* (20 January 1880), suggested something completely different. In contrast to the *Saturday Review*, it claimed that *Life and Habit* could have been written as a serious attempt at science rather than an ironical book. The review stated:

> The writer is, it appears, the author of a philosophical romance, entitled *Erewhon or, over the range*, which attracted some attention a year or two ago. As *Erewhon* is described in an advertisement on the flyleaf of this book as 'a work of satire and imagination' it may be as well to observe that Mr. Butler's latest production seems to have been written in perfect good faith. If the reader should therefore find himself laughing here and there, he should bear in mind that he laughs entirely upon his own responsibility.[70]

Looking at these reviews we can immediately notice how difficult it was for contemporary readers to label Butler's first book on science. The Victorian writer, of course, did not help the prospective readers. Butler, in the conclusion of the book, explained that *Life and Habit* was indeed written in good faith:

> Here, then, I leave my case, though well aware that I have crossed the threshold only of my subject. My work is of a tentative character, put before the public as a sketch or design for a, possibly, further endeavour, in which I hope to derive assistance from the criticisms which this present volume may elicit. Such as it is, however, for the present I must leave it.[71]

However, he also recognized that his first book on science was aimed at being a writing experiment. Butler was indeed aware of being a non-specialist writer, as he declared a few pages later:

> Of course, if I were a specialist writing a treatise or primer on such and such a point of detail, I admit that scientific accuracy would be de rigueur; but I have been trying to paint a picture rather than to make a diagram, and I claim the painter's license 'quidlibet audendi.' I have done my utmost to give the spirit of my subject, but if the letter interfered with the spirit, I have sacrificed it without remorse.[72]

There is one final example to look at in order to further explore the complexity of Butler's writing on science: 'God the Known and God the Unknown'. After the publication of *Life and Habit* and *Evolution, Old and New*, Butler decided to write a series of articles, eight in total, to be published in the periodical *The Examiner* in May, June, and July 1879, respectively.[73] Since 1808, *The Examiner* published articles which combined surveys of politics, literature, drama, and the arts. The periodical was aimed at providing Victorian readers with essays designed to explore in more detail certain topics or questions when compared to daily newspapers.[74] Victorians considered *The Examiner* a leading intellectual and radical journal. In other words: the perfect place where to explore key political, social, scientific, and religious issues. This made *The Examiner* the ideal periodical for Butler's desire to explore and discuss the main debates of his time.

In 'God the Known and God the Unknown', Butler investigated the different roles played by God in nature, from old pantheism to the new, and exciting, investigations of God's place in the evolutionary debate. It is important to clarify that these articles were neither intended to be a philosophical, literary, nor a theological examination of the question about the divine. As in *Erewhon* and *Life and Habit*, Butler approached the matter without relying on one specific methodology.

Butler's interest, speculative in nature, was to explain to readers how questions concerning God were still central to the science and philosophy of the late nineteenth century. However, despite the interesting central topic, there are two additional aspects that make these articles worth further historical attention. First, although the papers were designed to challenge religious/theological problems, they also look into philosophical and evolutionary questions in a similar manner to what was done previously in *Life and Habit*. Second, the use of different sources including religious documents and extracts from classic literature, alongside the discussion of contemporary philosophical and scientific texts, made 'God the Known and God the Unknown' another worthy case study to explore how Butler was unable to target one specific audience. Although the content of Butler's articles could appear to be quite clear in the eyes of an expert, it was foggy, vague, and even pretentious for the palate of an uneducated Victorian.

Before venturing into the content of the articles, there are some preliminary points to make. Although mostly discussed by theologians and philosophers before 1860, the question concerning the role and place of God in nature was still investigated and debated in the late 1880s.[75] Consequently, theological and spiritual questions were also still largely explored and discussed by the academic community in books, essays, and the periodical press.[76] Butler, of course, wanted to have a say in this debate. In doing so, his main interest, as already explained in *Life and Habit*, was not to engage with academics, but rather to open a dialogue with the general public. Butler clarifies the scope of the articles at the beginning of 'God the Known and God the Unknown' as:

> Firstly, I can demonstrate, perhaps more clearly than modern science is prepared to admit, that there does exist a single Being or Animator of all living things — a single Spirit, whom we cannot think of under any meaner name than God; and, secondly, I can show something more of the persona or bodily expression, mask, and mouthpiece of this vast Living Spirit than I know of as having been familiarly expressed elsewhere, or as being accessible to myself or others, though doubtless many works exist in which what I am going to say has been already said.[77]

Butler's analysis started by looking at God in the pantheistic tradition, which relied on the assumption that 'God is everything and everything is God', but after assessing the different iterations of pantheism throughout history, the Victorian thinker did not find this idea convincing. He explained that:

> we can see the gold-fish as forming one family, and therefore as in a way united to the personality of the parents from which they sprang, and therefore as members one of another, and therefore as forming a single growth of gold-fish, as boughs and buds unite to form a tree; but we cannot by any effort of the imagination introduce the bowl and the water into the personality, for we have never been accustomed to think of such things as living and personal.[78]

It is worth noting here how the language and examples used by Butler were designed to explain complex concepts by using very simple case studies, which could be easily understood by the general public. This is a device that Butler deployed throughout this series of articles. In 1918, Felix Grendon's 'Samuel Butler's God' explained how the Victorian writer produced a 'sermon on the brotherhood and Godhead of man without a line of preaching and without single mention of the words brother hood or Godhead'.[79] According to Grendon, Butler's aim was to write 'an essay in straightforward thinking and a matter of cool common-sense'.[80] Grendon's interpretation is correct but misses one key aspect of Butler's style of writing. Although Butler used a relatively simple writing style and straightforward thinking, the aim of 'God the Known and God the Unknown' was to also advance his biological theory. This intention is especially clear in the final article of the series.

Returning to the content of 'God the Known and God the Unknown': after dealing with pantheism, the Victorian writer turned to explore the significance of orthodox deism. For Butler, orthodox deism is based on the interpretation of God in his spiritual form, and therefore impersonal. For Butler, this interpretation of God has no value. He explains: 'no conception of God can have any value or meaning for us which does not involve his existence as an independent Living Person of ineffable wisdom and power, vastness, and duration both in the past and for the future.'[81] Therefore, Butler came to the conclusion that orthodox deism is merely a form of atheism. He writes:

> it must be remembered there can be no God who is not personal and material: and if personal, then, though inconceivably vast in comparison with man, still limited in space and time, and capable of making mistakes concerning his own interests, though as a general rule right in his estimates concerning them. Where, then, is this Being? He must be on earth, or what folly can be greater than speaking of him as a person? What are persons on any other earth to us, or we to them? He must have existed and be going to exist through all time, and he must have a tangible body. Where, then, is the body of this God? And what is the mystery of his Incarnation?[82]

The final section of the essay is the most interesting. Here, Butler presents a third possibility: to conceive of God as we conceive of hope (which we personify without the need of a real body). However, this final possibility, Butler recognizes, is not feasible because no conception of God could have any meaning for us unless it involved his existence as an independent material living being. Butler explains:

> I have repeatedly said that we ought to see all animal and vegetable life as uniting to form a single personality. I should perhaps explain this more fully, for the idea of a compound person is one which at first is not very easy to grasp, inasmuch as we are not conscious of any but our more superficial aspects, and have therefore until lately failed to understand that we are ourselves compound persons.[83]

Butler's argument here heavily relies on the one he previously advanced in *Life and Habit*. He directly cites his book and suggests that in order to make a proper attempt to understand the nature of God we should start from the assumption that: 'Each cell in the human body is now admitted by physiologists to be a person with an intelligent soul, differing from our own more complex soul in degree and not in kind.'[84] It is not surprising then to see how Butler arrives at the conclusion that 'memories which all living forms prove by their actions that they possess — the memories of their common identity with a single person in whom they meet — is incontestable proof of the fact of being animated by a common soul'.[85] Interestingly, Butler's idea of Lamarckian design subsequently becomes central to his investigation of the divine. In particular, the possibility of thinking about the evolution of a living being as part of a process that involves the organism and its ancestors as a whole becomes, for Butler, the key for creating a bridge between science and religion.

In this section of 'God the Known and God the Unknown', Butler shifts the argument, method, and style of writing from a theological examination of the divine to a scientific one. This change can be seen clearly in the essay when Butler writes:

> It is certain, therefore, that all living forms, whether animal or vegetable, are in reality one animal; we and the mosses being part of the same vast person in no figurative sense, but with as much bona fide literal truth as when we say that a man's finger-nails and his eyes are parts of the same man.[86]

Here, Butler presents an idea very similar to what he suggested in his first book on science. In *Life and Habit* Butler had already put forward the hypothesis that we are one person with our ancestors.[87] It follows, then, that in 'God the Known and God the Unknown', Butler realized that the only fruitful way for exploring and defining the nature and place of God is in relation to evolution.

This approach has a double value for Butler. First, the writer explains how the study of theology must embrace the new discoveries made by evolutionists in the nineteenth century. Second, Butler uses this new approach as an opportunity for self-promoting his own theory of evolution. He explains that in order to fully understand the nature of God, 'Evolution must be grasped, and not Evolution as taught in what is now commonly called Darwinism, but the old teleological Darwinism of eighty years ago.'[88] And just a few pages later, Butler further stresses the need to use science to define God and his work:

> THE reader will already have felt that the panzoistic conception of God — the conception, that is to say, of God as comprising all living units in His own single person — does not help us to understand the origin of matter, nor yet that of the primordial cell which has grown and unfolded itself into the present life of the world. How was the world rendered fit for the habitation of the first germ of Life? How came it to have air and water, without which nothing that we know of as living can exist? Was the world fashioned and furnished with aqueous and atmospheric adjuncts with a view to the requirements of the infant monad, and to his due development? If so, we have evidence of design, and if so of a designer, and if so there must be some far vaster Person who looms out behind our God, and who stands in the same relation to him as he to us.[89]

The questions, language, and ideas presented in the passage above are very similar to some of the questions posed in *Erewhon* and then *Life and Habit*, especially in relation to the origin of life and the role of design in evolution. Unfortunately, 'God the Known and God the Unknown' did not provide an answer to any of these questions. Butler simply left this matter open for discussion. Once again, Butler's desire to engage with two different audiences at once confused prospective readers who were unable to grasp the real aim of those articles.

To find a solution to this matter, Victorian readers had to wait for the publication of Butler's next book on science: *Unconscious Memory*. There, Butler reopened the question regarding the nature of God mixing once again scientific writing with his philosophical ideas concerning evolution and the mind. Butler realized that in 'God the Known and God the Unknown' he had committed some mistakes. So, in the concluding chapter of *Unconscious Memory* Butler explained, referring to his articles:

> I would recommend the reader to see every atom in the universe as living and able to feel and remember, but in a humble way. He must have life eternal as well as matter eternal; and the life and the matter must be joined together inseparably as body and soul to one another. Thus he will see God everywhere, not as those who repeat phrases conventionally, but as people who would have their words taken according to their most natural and legitimate meaning; and he will feel that the main difference between him and many of those who oppose him lies in the fact that whereas both he and they use the same language, his opponents only half mean what they say, while he means it entirely... We shall endeavour to see the so-called inorganic as living, in respect of the qualities it has in common with the organic, rather than the organic as non-living in respect of the qualities it has in common with the inorganic.[90]

Leaving aside theology, *Unconscious Memory* further stressed the necessity of thinking about the divine predominately in relation to evolution. However, once again, Butler did not provide any explanation but rather used this part of the book to strengthen his claim that evolution must be explained as a Lamarckian process. Instead of simply questioning the nature of God from a standard theological standpoint, Butler used his knowledge of science and evolution to shed new light on the matter. Nonetheless, despite his effort in trying to create a bridge between these two different debates, this endeavour made his work, once again, difficult to categorize. Both the professional and the lay readership failed to understand why this Victorian author tried to answer theological questions using the language, mechanism, and ideas offered by evolutionary science and psychology.

The Problem of Identity and the Reception of Butler's Work

As assessed by Clara Stillman in 1932, Butler's

> contradictory opinions, his excessive or insufficient emotions, his whimsical dislikes, his oddities of behaviour, his sensitiveness, his self-protective way of life were the tentative unconscious efforts at adjustment of a soul that had not received its full birth right of initiation in human relations.[91]

Butler was not only a problematic product of Victorian culture. He was a problematic Victorian. This was because, as Stillman explained, many of his ideas were ahead of his time:

> Butler was alone among English writers in thoroughly transcending the current philosophic and psychological assumptions of his period, and he is alive today because many of his ideas have become the current assumptions of ours.[92]

This was particularly the case with his science of the mind, which was not recognized as a proper theory until the very early twentieth century. However, his ideas and writing style did not fit with the professional turn that science was taking in Britain in the second part of the nineteenth century. It follows that Butler's ability to work across different genres and fields of knowledge made his reception extremely problematic.

Butler's writing on science would probably have been perfect for the cultural debate on evolution of the 1840s. As Grant Allen (1848–1899) explained in his review of *Evolution, Old and New* (*Examiner*, 17 May 1879), Butler approached evolution exactly like Jonathan Swift, at his time, critically discussing English society. He wrote: 'If Swift had lived to be an evolutionist, he would have written Mr Butler's new book.'[93] Nonetheless, Allen's comment implied another point. If Butler's *Evolution, Old and New* was considered the 'new' version of *Gulliver's Travels* (1726), his writing on science could not be considered scientific or serious at all.

Butler's problem was then a matter of identity and authority. He was neither considered a novelist nor a scientist, and this was due to his unclassifiable working activity. Butler was active in both literature and science until his death in 1903. The unclassifiable nature of Butler's work makes us pose some fundamental questions: was Butler a writer or a scientist, a traveller or an artist — or all of these? And how did being a polymath influence his credibility and reputation? These are the questions that must be answered in the next chapters where Butler's engagement with professional science and the consequent decline of his fame will be explored in detail. For now, the only preliminary definition that can be made is that Butler was, in his own right, a polymath working in between professionals and amateurs, and the reception of his work was, both positively and negatively, the result of that.

Notes to Chapter 2

1. Samuel Butler, 'God the Known and God the Unknown', *The Examiner* (London, 24 and 31 May; 14, 21, 28 June; 12, 19, 26 July 1889); also published in pamphlet form, with a note by R. A. Streatfeild (1909). In this chapter, I will cite the following edition: Samuel Butler, *God the Known and God the Unknown* (New Haven: Yale University Press, 1917), here p. 71.
2. George Bernard Shaw, *Major Barbara* (London: The Court Theatre, 1907), pp. i–iii.
3. See Paradis, *Samuel Butler, Victorian against the Grain*, pp. 3–20 (esp. p. 5).
4. Christina Myers-Shaffer, *The Principles of Literature: A Guide for Readers and Writers* (New York: Barron's Educational Series, 2000), p. v.
5. Myers-Shaffer, *The Principles of Literature*, pp. 9–10.
6. Peter Raby, *Samuel Butler: A Biography* (London: Hogarth Press, 1991), p. 116.
7. Raby, *Samuel Butler: A Biography*, pp. 161–65.
8. Philip Cohen, 'Stamped on His Work: The Decline of Butler's Literary Reputation', *The Journal of the Midwest Modern Language Association*, 18.1 (1985), 65–66.
9. See Paradis, *Samuel Butler, Victorian against the Grain*, p. 5.
10. Lee Holt, 'Samuel Butler up to Date', *English Literature in Transition, 1880–1920*, 3.1, (1960), 17–21.
11. Butler, *The Notebooks*, p. 106.
12. From the early scholarship in the 1920s onwards, several critical studies have been published on Samuel Butler's work. Although the majority of existing studies look at his work within the Victorian literary debate and recognize the importance of his contribution, others, especially the ones published longer ago, rate Butler poorly as just a satirical novelist and thinker without giving any credit to his work (see Irvine, *Apes, Angels and Victorianism*, p. 246). However, other accounts — including George Bernard Shaw's *Major Barbara* (1907) and the more recent biographical account written by Raby (1991) — have recognized the significance of Butler's fictional writings.
13. As discussed in the introduction, the negative reception of Butler's work has changed in recent years: scholars have started to rethink his role in the Victorian evolutionary debate as a popularizer of science, a forgotten pioneer of psychology, and even as a philosopher of the unconscious.
14. Malcolm Muggeridge, *The Earnest Atheist: A Study of Samuel Butler* (London: Eyre & Spottiswoode, 1936).
15. Muggeridge, *The Earnest Atheist*, p. ix.
16. Muggeridge, *The Earnest Atheist*, p. vii.
17. Muggeridge, *The Earnest Atheist*, p. vii.
18. Irvine, *Apes, Angels and Victorianism*, p. 246.
19. Irvine, *Apes, Angels and Victorianism*, p. 249.
20. Similarly, in Peter Bowler's *Evolution: The History of an Idea*, Butler's science is dismissed as the attempt of an amateur without any chance to be really successful. Bowler's reading of Butler's work fits within the tradition, which did not recognize the significance of Butler's Lamarckism and science of the mind. See, Bowler, *Evolution: The History of an Idea*, p. 259.
21. Shaffer, *Erewhons of the Eye*, p. xi.
22. Shaffer, *Erewhons of the Eye*, p. xi.
23. Paradis, 'Introduction', in *Samuel Butler, Victorian against the Grain*, ed. by Paradis, pp. 3–18 (p. 3).

24. Paradis, *Samuel Butler, Victorian against the Grain*, ed. by Paradis, p. 7.
25. Gillian Beer, *Open Fields: Science in Cultural Encounter* (Oxford: Clarendon Press, 1996), pp. 173–95.
26. James Secord, 'Knowledge in Transit', *Isis*, 95 (2004), 654–72.
27. Beer, *Darwin's Plots*, pp. 53–60, pp. 65–70.
28. Beer, *Darwin's Plots*, p. 5.
29. James Secord, *Victorian Sensation: The Extraordinary Publication, Reception, and Secret Authorship of 'Vestiges of the Natural History of Creation'* (Chicago: University of Chicago Press, 2000), pp. 4–5.
30. Butler, *The Notebooks*, p. 109.
31. Butler, *The Notebooks*, p. 119.
32. Samuel Butler, *The Family Letters of Samuel Butler, 1841–1866* ed. by Arnold Silver (Palo Alto: Stanford University Press, 1962), p. 118.
33. This figure can be found in the section 'Analysis of the Sales of my Books' included in Butler, *The Notebooks*, p. 368. As the reader will notice, the *The Way of all Flesh* is not included in the list. This is because, although *The Way of all Flesh* was Butler's most successful novel (both nationally and internationally), it was published only after his death in 1903.
34. Beer, *Open Fields: Science in Cultural Encounter*, p. 78.
35. Sue Zemka, '*Erewhon* and the End of Utopian Humanism', *ELH*, 69.2 (2002), 339–49.
36. Gillian Beer, 'Butler, Memory, and the Future', in Samuel Butler, *Victorian against the Grain*, ed. by Paradis, pp. 45–57.
37. Beer, 'Butler, Memory, and the Future', p. 55.
38. Beer, 'Butler, Memory, and the Future', pp. 55–57.
39. Samuel Butler, *Erewhon, or, Over the Range* (London: Fifield, 1908), p. 237.
40. Butler, *Erewhon*, p. 251.
41. Thomson, William, *Mathematical and Physical Papers* (Cambridge: Cambridge University Press, 2011), p. 188.
42. Butler, *Erewhon*, p. 319.
43. Butler, *Erewhon*, p. 272.
44. Joshua A. Gooch, 'Figures of Nineteenth-Century Biopower in Samuel Butler's Erewhon', *Nineteenth-Century Contexts: An Interdisciplinary Journal*, 36.1 (2014), 53–71 (p. 53).
45. Butler, *Erewhon*, p. 251.
46. Butler, *Erewhon*, p.251
47. Butler, *The Notebooks*, p. 44.
48. Butler, *The Notebooks*, p. 46.
49. Butler, *The Notebooks*, p. 50.
50. Butler, *The Notebooks*, p. 50.
51. Butler, *The Notebooks*, p. 50.
52. A good account of the work of Lamarck is provided in Jablonka and Lamb, *Epigenetic Inheritance and Evolution*.
53. Lamarck, *Philosophical Zoology*, p. 113.
54. Butler, *The Notebooks*, p. 48.
55. Butler, *Erewhon*, pp. 294–95.
56. During his life, Butler published two different editions of the novel. The place and significance of Lamarckian ideas, in the novel, changed significantly between the first and second editions. The first edition of *Erewhon* (1872) was written as a tribute to Darwin's theory of evolution. The revised edition of 1901 expanded the original story by adding two new chapters and a preface. More importantly, this edition embodied Butler's revised view of evolution, which had shifted markedly away from Darwin's and towards Lamarck's. I have discussed the differences between these two editions of the novel here: Cristiano Turbil, 'Memory, Heredity and Machines: From Darwinism to Lamarckism in Samuel Butler's *Erewhon*', *Journal of Victorian Culture* (2019) <https://doi.org/10.1093/jvcult/vcz038> [accessed 19.11.2019].
57. Butler, *Erewhon*, p. 293.
58. Butler, *Erewhon*, p. 293.
59. Butler, *Erewhon*, p. vii.

60. Butler, *Erewhon*, p. xi.
61. Butler, *Life and Habit*, pp. viii–ix.
62. In 1880, at the very beginning of *Unconscious Memory*, Butler explained how he became one of the many admirers of Darwin's work: 'As a member of the general public, at that time residing eighteen miles from the nearest human habitation, and three days' journey on horseback from a bookseller's shop, I became one of Mr. Darwin's many enthusiastic admirers, and wrote a philosophic dialogue (the most offensive form, except poetry and books of travel into supposed unknown countries, that even literature can assume) upon the Origin of Species. This production appeared in the *Press*, Canterbury, New Zealand, in 1861 or 1862, but I have long lost the only copy I had.' See Butler, *Unconscious Memory*, p. 11.
63. Butler, *The Notebooks*, pp. 40–41.
64. Butler, *Erewhon*, p. xi
65. Butler, *Erewhon*, p. xi.
66. Butler, *The Notebooks*, p. 53.
67. Butler, *Erewhon*, p. 53.
68. Butler, *Life and Habit*, pp. 1–2.
69. Review of Butler, *Life and Habit*, in *The Saturday Review* (26 December 1878), 119–21 (p. 119).
70. Review of Samuel Butler, *Life and Habit*, in *Daily News* (20 January 1880), [n.p.].
71. Butler, *Life and Habit*, p. 294.
72. Butler, *Life and Habit*, pp. 301–02.
73. The articles were then republished as a book in 1909 by the editor A. C. Fifield, with a prefatory note by R. A. Streatfeild.
74. *The Cambridge History of English and American Literature: An Encyclopaedia in Eighteen Volumes*, 18 vols, ed. by A. W. Ward and others (Cambridge: Cambridge University Press, 1907–2000), XII: *The Romantic Revival*, ed. by A. W. Ward and A. R. Waller (1915), pp. 244–48.
75. The study of theology changed significantly during the Victorian period. For instance, James Russell Perkin's *Theology and the Victorian Novel* identifies that religion was a central part of Victorian culture and had a considerable impact on literature not only in those novels in which religion was a primary issue. Especially in relation to natural history, theology represented the main detractor of natural selection. The study of theology was used in justifying the necessity of rethinking evolution in metaphysical terms and not only materialistically as done by Darwin. To know more about these issues, see James Russell Perkin, *Theology and the Victorian Novel* (Montreal: McGill-Queen's University Press, 2009), pp. 5–8. In addition, theology was also still central to university teaching and governance. For instance, the teaching of theology was defined by John Henry Newman, in *The Idea of a University* as: 'In a word, Religious Truth, is not only a portion, but also a condition of general knowledge.' See John Henry Newman, *The Idea of a University: Defined and Illustrated* (London: Pickering, 1873), p. 57.
76. Darren J. N. Middleton 'James Russell Perkin, Theology and the Victorian Novel', *Religious Studies Review*, 36.4 (2010), 289.
77. Butler, *God the Known and God the Unknown*, p. 19.
78. Butler, *God the Known and God the Unknown*, p. 26.
79. Felix Grendon, 'Samuel Butler's God', *The North American Review*, 208 (1918), 277–86 (p. 277).
80. Grendon, 'Samuel Butler's God', p. 277.
81. Butler, *God the Known and God the Unknown*, p. 42.
82. Butler, *God the Known and God the Unknown*, pp. 51–52.
83. Butler, *God the Known and God the Unknown*, p. 55.
84. Butler, *God the Known and God the Unknown*, p. 55.
85. Butler, *God the Known and God the Unknown*, p. 62.
86. Butler, *God the Known and God the Unknown*, p. 62.
87. The idea that we are one person with our ancestors was then discussed again, in *Unconscious Memory* (p. 8), where Butler writes: 'It follows from this that all living animals and vegetables — being as appears likely, if the theory of evolution is accepted — descended from a common ancestor, are in reality one person and united to form a body corporate of whose existence, however, they are unconscious. There is an obvious analogy between this and the manner in

which the component cells of our bodies unite to form one single individuality, of which it is not likely they have a conception, and with which they have probably only the same partial and imperfect sympathy as we, the body corporate, have with them.'
88. Butler, *God the Known and God the Unknown*, p. 71.
89. Butler, *God the Known and God the Unknown*, p. 83.
90. Butler, *God the Known and God the Unknown*, p. 45.
91. Clara Stillman, *Samuel Butler: A Mid-Victorian Modern* (New York: The Viking Press, 1932), p. 11.
92. Stillman, *Samuel Butler*, pp. 7–8.
93. Grant Allen, review of Butler, *Evolution, Old and New*, in *The Examiner* (17 May 1879), 646–47.

CHAPTER 3

The Rise and Fall of Butler's Fame

> Mr Samuel Butler was a humourist, and he did many things, handicapping himself thus doubly, as he well knew, to the world's view. The author of *Erewhon* was not supposed to be serious in anything, and in this age of specialism for a classic scholar to know anything of art or science is considered indecent, if not inconceivable.
>
> The Athenaeum, 28 June 1902[1]

Since Butler moved to New Zealand in the early 1860s, it was clear that he did not have a specific career in mind. Returning to London after a few years, Butler first tried to become a novelist, then, after publishing *Erewhon*, he quickly realized that his future was in painting. In the same period, he wrote his four books on science, had a quarrel with Darwin, and started attacking the work of other professional scientists. This attitude combined with the inability to work 'professionally' in just one field completely destroyed his credibility among both the Victorian scientific and lay communities. The decrease in Butler's fame can be seen by simply looking at the reception of his various books which, excluding *Erewhon*, did not sell well during his life.

In order to shed some light on the reasons why this Victorian author did not acquire as much fame as he might have done during his lifetime, the best place to start is Paradis' essay 'Butler after Butler: The Man of Letters as Outsider'. Paradis' work provides us with a detailed account of the reception of Butler's work and ideas in his own country immediately after his death to trace how the Victorian writer gradually became an outsider.[2] Paradis explains that in order to understand Butler's writing, we must think about 'the developments in Butler's work that contributed to his alienation from his Victorian audience and the manner in which he incorporated this alienation into a productive outsider persona that unified his work and enabled him to continue writing'.[3] My investigation here, which starts with a look at the sales of Butler's books, aims to expand on the work of Paradis by questioning and exploring the decline of his fame and the reasons behind it. My claim is that the Victorian tendency of considering Butler as an eccentric thinker directly shaped his reputation and the reception of his work and that, in turn, his attitude to go 'against the grain' made his work unacceptable to the people of his generation.

The negative reception of Butler's written work can be clearly seen in the section 'Analysis of the Sales of my Books' recorded by the author in his notebooks. Here,

Butler provides an account of the sales numbers of his books and, unfortunately, the consequent loss of money. Recorded by Butler in 1898 and then included in the printed version of the *Notebooks* edited by Henry Festing Jones in 1912, this report is the only official documentation authorized by the author. In looking at it, the first thing to note is a substantial reduction in the number of sales after the publication of *Evolution, Old and New*. This was due to a variety of different factors, which will be explored in detail in the next few pages. For now, one must look at what Butler himself wrote about this negative sales report:

> It will be noted that my public appears to be a declining one; I attribute this to the long course of practical boycott to which I have been subjected for so many years, or, if not boycott, of sneer, snarl and misrepresentation. I cannot help it, nor if the truth were known, am I at any pains to try to do so.[4]

This boycott, Butler explained, came from professional scientists, in particular, Darwin and his circle. The reduction in the sales of his scientific books also compromised the general selling of other publications including the ones on art and travelling. Published in 1897, *The Authoress of the Odyssey* sold only 165 copies, which signified a huge reduction of interest in Butler's work from the Victorian readership.

In order to understand why Butler became a Victorian outsider, in this chapter I will look at how his relationship with Darwin, his family, and his intellectual and professional circles influenced the rise and fall of his engagement with evolution and science. In particular, my aim is to look at the relationship between Butler and Darwin both before and after their famous quarrel. Only in this way can we shed some additional light on the role this quarrel played in the lessening of Butler's fame.

The Reception of Samuel Butler in England and the Quarrel with Charles Darwin

Peter Raby, in his biography of Butler, defines the Victorian writer as 'the archetypal rebel', who stood deliberately against the traditions of his time.[5] Raby's definition of the writer as a Victorian rebel, similar to the more recent epithet used by Paradis, 'Victorian against the Grain', fits with a new tradition of studies which presents Butler's work as a critical attempt to expose some of the weaknesses of the culture and science of his time. Although I agree with these definitions of Butler's character, I also think that in order to fully understand the reasons behind his attitude we need to look more broadly at why he was so critical and how this approach influenced the reception of his ideas.

It follows, that in order to reframe the reception of Butler's writings on science in Victorian England one must look at a variety of different aspects: from his attitude towards writing to his pride and personal relationships with his Victorian peers. Only in this way will it be possible to establish how the reception of Butler's work, by the people of his generation, determined the consequent rise and decline of his fame.

A good source of information to see how significant the quarrel with Darwin was for the reputation of Butler among Victorians are his obituaries. Newspaper and periodical articles about Butler's demise appeared in several British newspapers and periodicals such as *The Times*, *The Athenaeum*, *The Monthly Review*, and *The Eagle*. Several short articles were also published abroad, especially in Italian journals such as *Il Corriere Valsesiano*, *Il Monte Rosa*, *Quo Vadis?*, and *Nuova Antologia*.[6] A preliminary comparison between the English and Italian sources immediately shows a disparity in reception. In England, Butler's death was generally reported in a disinterested way. He was just the Victorian who wrote the novel *Erewhon* and had a quarrel with Darwin. In contrast, in Italy, as will be discussed later, Butler was described as a friend of the country and even as one of the key intellectuals of his generation. Butler received awards which celebrated his contribution to Italian culture, and was, from the north to the south, recognized and treated as a friend. So, it follows that we must first question why Butler's reception in his own country was so negative when compared to Italy and, second, we should see whether the decrease in Butler's fame was due to the quarrel he had with Darwin, which was particularly influential in England.

In looking at articles about Butler's death, two specific details come immediately to our attention. First, all of the obituaries agreed in defining Butler as a novelist (the author of the famous *Erewhon*) often without recognizing his contribution to the scientific debate of his time. Second, all of the newspapers referred to Butler's involvement in the quarrel with Darwin as one of the most relevant events of his life. For instance, in *The Times* (20 June 1903) we read:

> We regret to announce that Mr Samuel Butler, best known to his countrymen as the author of *Erewhon*, died on Wednesday night, in his 67th year. He was a remarkably gifted man, though he never won the success and recognition to which his intellectual force and his powers of expression entitled him. [...] he used to boast that his grandfather the Bishop attacked Darwin's Grandfather, that his father has been in controversy with Darwin's father, and he seemed to his own hostility to Darwin's system and his vindication of Lamarck as instances of a hereditary feud.[7]

Just a few lines later, the editor of *The Times* specifically emphasized Butler's anti-Darwinian feelings:

> It is certain that he set the greatest value on his anti-Darwinism writings, a judgment in which few impartial critics will agree. There was much that was stimulating in his books connected with this subject [...] but they were lacking in clearness, coherence and consistency, and their effect was inevitably imperfect and evanescent.[8]

It is easy to see from the extracts above how Butler was considered a gifted individual but also an individual unable to escape from his own ego and the animosity he developed about Darwin and his work. Indeed, among British men of science Butler's books were found to be stimulating on a purely speculative level but, as stated by *The Times*, 'lacking in clearness, coherence and consistency'.

This trend of considering Butler's work as an attempt to undermine Darwin's ideas can also be found in another obituary published in *The Athenaeum* by the journalist Vernon Horace Rendall. The general tone of the obituary is very similar to that of *The Times*, but here, Rendall dismissed Butler's personality and scientific work even further. Rendall made two points that are worth discussing. First, he underestimated Butler's work and contribution to the Victorian scientific and cultural debates by saying that he was just 'a humourist, and he did many things, handicapping himself thus doubly, as he well knew, to the world's view'.[9] Second, Rendall insisted in stressing how the Victorian writer 'was not supposed to be serious in anything, and in this age of specialism for a classic scholar to know anything of art or science is considered indecent, if not inconceivable'.[10] The criticism became even harsher when Rendall focused on Butler's writing on science:

> His scientific books [...] are now, perhaps, forgotten, and difficult to get, as only small editions were printed. I do not think that he considered them the best of his work, though the increasing body of Neo-Lamarckians might find them useful. The hereditary quarrel with Darwin and Darwin's forbears of which he used to speak was in later life at any rate not more than a jest, though he always felt that Darwin had not treated him quite fairly.[11]

Interestingly, although Rendall underestimated the importance of Butler's books on science, he recognized the increasing interest in neo-Lamarckism within the continental debate. Nonetheless, especially in the second part of the extract, it is clear that any possible interest in Butler's theory of evolution was overshadowed by the quarrel with Darwin.

Looking, in particular, at his writing on science, it is impossible to not notice how Butler's public and private quarrel with Darwin interfered in the reception of his work. The quarrel has been for a long time the primary historical source for understanding the relationship between the father of evolution and the author of *Erewhon*. Indeed, most of the definitive historical works on Butler and Darwin refer to their dispute as the key event that determined Butler's career. From the classic Basil Willey's *Darwin and Butler: Two Versions of Evolution* (1960) to the recent essay 'The Butler-Darwin Biographical Controversy in the Victorian Periodical Press' by Paradis (2004), the quarrel has been used by scholars to explore and question why Butler attacked Darwin and Darwinism.[12]

The famous quarrel began after the publication of Butler's second book on science, *Evolution, Old and New* (March 1879), and it continued until Darwin's death in 1882. At the heart of the quarrel lay a simple misunderstanding created by the forgotten acknowledgement of Butler's *Evolution, Old and New* in a publication about the life and work of Erasmus Darwin. Butler expected to find a reference to his work in Charles Darwin's preface to the English translation of the biography of Erasmus Darwin by E. Krause, entitled *Life of Erasmus Darwin* (1879).[13] No mention of Butler's *Evolution, Old and New* or work in general, was, however, included in the book, and the Victorian writer perceived this omission as a deliberate attempt to obscure his work.

One of the most complete historical accounts of the quarrel can be found in an appendix to *The Autobiography of Charles Darwin, 1809–1882*, edited by his granddaughter Nora Barlow. Barlow motivated her choice to include the quarrel in the autobiography with the following words:

> Today the once notorious quarrel between Samuel Butler and Charles Darwin is almost forgotten, and the short account in the complete version of the Autobiography, — printed here for the first time, — will only raise vague memories in the minds of most readers.[14]

In 1958, the quarrel between Butler and Darwin was long forgotten and discussed uniquely by those interested in Butler's work. However, Barlow's decision to look back at this old Victorian intellectual dispute, although original and innovative at the time, lacked much in terms of the number of sources analysed and the general contextualization of the events discussed. For instance, Barlow made reference only to a small number of sources, specifically a Cambridge dossier (a collection of letters stored in the manuscript room at Cambridge university library) and a pamphlet saying 'Henry Festing Jones, Butler's biographer and friend, brought out a Pamphlet, in 1911, now out of print, entitled "Charles Darwin and Samuel Butler, A Step Toward Reconciliation"'.[15]

In addition, as Barlow's aim was, of course, to shed some light upon this Victorian quarrel, the appendix included extracts from the correspondence between the two Victorians. However, Barlow focused only on the letters written after the publication of *Evolution, Old and New* in 1879. It follows that although useful in offering an overview of the events during and after the quarrel, Barlow's work did not fully discuss the relationship between Butler and Darwin. This general lack of context can be seen by looking at her opinion of Butler. Barlow wrote:

> Butler stands as the perpetual revolutionary, who only turned against Darwin after Darwin had become the acknowledged prophet. Darwin was rebelling against current biological concepts and delivered Man into the evolutionary machine; he rejected all easy speculators as ephemeral, and to him Butler and his theories remained ephemeral.[16]

Barlow described Butler as a sort of villain 'rebelling' against her grandfather. It does not surprise, then, to see how Butler's scientific ideas were presented and discussed not as a potential alternative to Darwin's theory but rather as an obvious attempt to bring into disrepute the revolution introduced into the biological debate by the *Origin of Species*. This point is well summarized in the last part of Barlow's appendix.

> Butler's satirical genius lashed the shams and hypocrisies of his time. His writings on quasi-scientific themes as well as his philosophy on the art of living, were based on his inward experience, in revolt against fact-finding materialism. In Natural Selection and its dependence on chance variation for its effectiveness, — though Darwin himself vacillated on this point as Butler very well knew, — Butler saw a complete surrender to a mechanical world, with Man as the supreme machine, and all effect of Mind and its striving ruled out as a guiding force. He formed his theory of Mind and Memory in the speculative manner of the previous century, following and extending the ideas of Lamarck and Dr. Erasmus Darwin, with acknowledged indebtedness to his own contemporary, Dr. Hering. Butler paraded the old theories in a new guise,

and took on the role of the maltreated, posthumous "enfant terrible" of the Physico-theologians of the 18th century. Butler's intervention into the scientific fold with this hybrid of science and philosophy could not be tolerated by the new biological school of Darwin and Huxley.[17]

Interestingly, Barlow recognized some of the key aspects about Butler's science of the mind, namely: his speculative approach, the use of Lamarckism, and his strong criticism of the purely mechanical interpretation of evolution proposed by Darwin. However, for Barlow, Butler's work was also an attempt to present old theories which could not have been tolerated or discussed next to the works of Darwin and Huxley. This was because Butler's old-fashioned understanding of evolution was stuck in a moment where science was still a matter of speculative inquiry and did not embrace, as I will discuss in the next chapter, the professional shift of the late nineteenth century. In addition, because the writer rejected so rigorously Darwin's materialism, his work on the 'science' of the mind was simply perceived as pseudoscientific nonsense. It is therefore easy to see why Barlow's appendix on the quarrel focused predominantly on a personal/family level rather than discussing the key differences between their ideas of evolution.

To see a shift in this interpretation of the quarrel we need to wait over fifty years for the publication of Paradis' essay 'The Butler-Darwin Biographical Controversy', which provides a completely different explanation about the significance of the quarrel in the Victorian period.[18] In contrast to the claim made by Barlow, Paradis states that the quarrel was more than a family dispute. For the historian, the main outcome of the argument was the dismissal of Butler from the Victorian scientific community and the total rejection of his work on science by the Victorian press.[19]

Paradis' paper presents two interpretations of the quarrel that were overlooked by Barlow. First, in the Butler-Darwin controversy, the main aspect to look at is not the private correspondence between the two Victorians but rather the more public dimension of the quarrel in the periodical press.[20] Second, most of Butler's detractors criticized his persona and his personal attack on Darwin rather than the significance of his theory of evolution.[21] Thus, Butler became the villain, the anti-Victorian, and the nemesis of Victorian science.

In order to fully understand the role that Charles Darwin's theory of evolution had in the development of Butler's work and desire to contribute to the scientific debate of his time, one must start with the early 1860s, long before the famous quarrel. In these years, Butler's literary and scientific writings were recognized with enthusiasm by Darwin and his circles. There are a considerable number of letters between Butler, Charles Darwin, and Francis Darwin in the archive of the University of Cambridge Library.[22] This material offers an overview of the relationship between these Victorians and tells a different story concerning the real relationship between Butler, Darwin, and their respective families.

Before the publication of *Evolution, Old and New* in 1879, Butler was on friendly terms with the whole Darwin family. He used to visit Charles Darwin at Down House and to meet with his son Francis ('Frank') in London on a relatively regular basis to discuss art, literature, and science.[23] Henry Festing Jones' *Memoir* reports that Butler was also known by and even on friendly terms with, Mivart, Alfred Russel Wallace, Joseph Dalton Hooker (1817–1911), Grant Allen, and many other scientific personalities of the period.[24] It is therefore important to think about the relationship between Butler and Darwin as a sort of private Victorian family tale, which started long before 1879. However, the quarrel was also public and undermined the significance and consequent reception of Butler's work especially among those professionals close to Darwin.

In order to reconstruct the history between Butler's and Darwin's families, it is worth exploring what Francis Darwin wrote about their relationship in his *Life and Letters of Charles Darwin*:

> The friendship between the families of Darwin and Butler began many years ago. Charles Darwin's father, Robert, was the leading doctor in Shrewsbury when Butler's grandfather, Dr. Butler, was headmaster of Shrewsbury School. Charles Darwin and Butler's father, Canon Butler, were schoolfellows at Shrewsbury, under Dr. Butler, and undergraduates together at Cambridge. They spent the summer of 1828 together on a reading-party at Barmouth, and Canon Butler said of Charles Darwin, 'He inoculated me with a taste for Botany which has stuck by me all my life'.[25]

In regard to Samuel Butler, the first reference to his work can be found in a series of letters written by Charles Darwin in the early 1860s concerning a 'New Zealand farmer' with a genuine passion for evolution. As discussed in chapter two, in 1862, Butler published under a pseudonym the essay 'Darwin on the Origin of Species: A Dialogue', in the periodical *The Press* of Christchurch, New Zealand. The dialogue was extremely well received by the colonial community and created a large ripple that even reached Charles Darwin back in England. This is confirmed in a letter Darwin wrote about the dialogue:

> Down. | Bromley. | Kent. S.E.
> March 24th
>
> *Private*
>
> M^r. Darwin takes the liberty to send by this post to the Editor a New Zealand newspaper for the very improbable chance of the Editor having some time spare space to reprint a Dialogue on Species. This Dialogue, written by some quite unknown to M^r. Darwin, is remarkable from its spirit & from giving so clear & accurate a view of Mr. D^s. theory. It is, also, remarkable from being published in a Colony exactly 12 years old, in which, it might have thought, only material interests would have been regarded. —
>
> Yours Obediently | Ch. Darwin[26]

Darwin expressed an interest in the 'Dialogue on Species' for two main reasons. First, Darwin appreciated its spirit and accuracy in explaining natural selection. Second, the 'Dialogue' had particular relevance as it was written in New Zealand, a young colony where it was surprising to see how Darwin's work was received and discussed with such a genuine interest so far away from Britain.[27]

Darwin's fascination with Butler's 'Dialogue' grew with time. In another letter (18 July 1863), this time to John (Julius) Haast (1822–1887), a German geologist working in New Zealand,[28] Darwin spoke about a mysterious dialogue, saying: 'I wonder whether you were the Author of a very amusing & really excellently done Dialogue on Natural Selection, in a New Zealand paper, which was sent to me?'[29] For the whole year, the secret concerning the author of the dialogue remained hidden while Darwin's interest kept on growing. The solution of this mystery arrived only a few months later when in a letter from Darwin's wife Emma to Joseph Hooker, dated 7 December 1863, we read: 'Also 2 squibs by the Author of the Dialogue in the New Zealand paper on Origin. He is a Mr Butler Grandson of the old master of Shrewsbury C.'s schoolmaster.'[30]

It is worth dedicating a few additional words to John (Julius) Haast. Haast — besides being one of Butler's friends in New Zealand and the person who first established a link between Butler and Darwin — was also an internationally recognized geologist.[31] Haast became one of the first professors of geology at Canterbury University College and curator of the Canterbury Museum. On Butler, Haast wrote, in his 1879 book *Geology of the Provinces of Canterbury and Westland*:

> We followed this opening to the Rangitata, having the snow-covered peaks of the central range before us; and, after descending several hundred feet into the bed of the river Potts where it joins the Rangitata, we crossed that river and reached Mesopotamia, then the sheep station of Mr Samuel Butler, where I established my head-quarters.[32]

The fact that Haast established his headquarters in Butler's sheep station is very important and it can tell us a lot about Butler's engagement with science while in the colony. Having an actual scientist to talk with offered the young intellectual an opportunity to look at science from the inside and critically discuss some of the key topics of the period: above all, evolution. Haast was extremely important in the development of Butler's critical understanding of science both in New Zealand and after his return to London. Indeed, the friendship with Haast continued until the death of the geologist in 1887. Butler always kept his German friend informed about his scientific work.[33]

The next phase of our story occurred in 1865. Butler's name and work are now known and appreciated by Darwin. The main subject of this group of letters is Butler's new work: *The Evidence of the Resurrection of Christ* (1865). The pamphlet analysed the accounts of the crucifixion and burial of Jesus provided by the Gospels, putting forward the theory that the Resurrection was no miracle but rather the result of Jesus losing and later regaining consciousness. As previously done with the 'Dialogue', Butler's aim was to discuss a serious matter but in a tone that could keep interested the professional readership and the general public alike. The *Evidence of the Resurrection of Christ* was sent to Darwin in the summer of 1865 alongside a note regarding a possible return of the writer to London. On 30 September 1865, Darwin wrote to Butler:

> I am much obliged to you for so kindly sending me your 'Evidence &c — '
> We have read it with much interest. It seems to me written with much force, vigour, & clearness; & the main argument is to me quite new. I particularly agree with all you say in your preface. I do not know whether you intend to

return to New Zealand & if you are inclined to write I should much like to know what your future plans are.[34]

Butler replied on 1 October, with a long letter that highlighted two important points. First, Butler discussed his plans for the future, mentioning a possible return to London and the desire to become an artist. Second, and more importantly, Butler returned to his dialogue on evolution to clarify some aspects concerning his use of Darwin's theory of natural selection:

> I always delighted in your origin of species as soon as I saw it out in N.Z. — not as knowing anything whatsoever of natural history, but it enters into so many deeply interesting questions, or rather it suggests so many that it thoroughly fascinated me. I therefore feel all the greater pleasure that my pamphlet should please you however full of errors it may be.
> The first dialogue on the origin which I wrote in the Press called forth a contemptuous rejoinder from (I believe) the Bishop of Wellington — (please do not mention the name, though I think that at this distance of space & time I might mention it to yourself) I answered it with the enclosed which may amuse you. I assumed another character because my dialogue was in my hearing very severely criticised by two or three whose opinion I thought worth having, and I deferred to their judgement in my next. I do not think I should do so now. I fear you will be shocked at an appeal to the periodicals mentioned in my letter, but they form a very staple article of bush diet, and we used to get a good deal of superficial knowledge out of them. I feared to go in too heavy on the side of the origin because I thought that having said my say as well as I could I had better now take a less impassioned tone: but I was really exceedingly angry.[35]

Butler's aim was to clarify some of the criticism his work had received in New Zealand focusing in particular on the dispute he had with the Bishop of Wellington. The letter above, in line with his work on the 'Resurrection', highlights how the author shared with Darwin the belief that an antagonism between evolution and religion was inevitable.

There is another letter that needs to be examined in order to further understand the significance of Butler's engagement with science and evolution in New Zealand. On 26 December 1865, Darwin wrote to Haast with some updates about Butler:

> Mr S. Butler is now established in London as an artist. He lately sent me a clever theological pamphlet. I should have much liked to have seen him here & have heard N.Z. news, but the bad state of my health has rendered this impossible.[36]

There is one final group of letters, written before the beginning of the quarrel and regarding Butler's scientific writings, that must be explored here: In the period between the publication of *Erewhon* (1872) and *Life and Habit* (1878), we can see an increase in the number of letters exchanged between Butler and Darwin, which show a growing mutual respect between the two. As I have discussed in the previous chapter, when Butler published *Erewhon*, reviewers were unable to fully understand the engagement with Darwin's work in the novel. The *Athenaeum*, for instance, defined *Erewhon* as 'an attempt to reduce to the absurd the whole theory of evolution'.[37] The *Fortnightly Review* summarized the book as an example of satire of unmistakable ingenuity.[38] Similarly, the *British Quarterly Review* defined the science in *Erewhon* as 'intended as a satire' and dismissed Butler's claims by saying:

> After reading the book with some care, we understand generally that its aim is to show the absurdity and inconsistency of certain current opinions in religion, science, and social life; but we are utterly at loss to see the relevancy or meaning of many of the illustrations selected.[39]

On 11 May 1872, Butler wrote a letter to Darwin to explain the misunderstanding generated by his satirical reading of Darwin's natural selection in *Erewhon*. It read:

> I venture about the liberty of writing to you about a portion of the little book 'Erewhon' which I lately published, and which I am afraid has been a good deal misunderstood. I refer to the chapter upon Machines in which I have developed and worked out the obviously absurd theory that they are about to supplant the human race and be developed into a higher kind of life. [...] I therefore thought it unnecessary to give any disclaimer of an intention of being disrespectful to the Origin of Species, a book for which I can never be sufficiently grateful, though I am well aware how utterly incapable I am of forming any opinion on a scientific subject which is worth a moment's consideration.[40]

There is no record of any reply to this letter. However, Butler was invited to visit Darwin and his family at Down House to discuss the content of his controversial novel.[41] A few weeks later, on 30 May 1872, Butler wrote again to Darwin proposing to send to him a copy of the second edition of *Erewhon* (*Darwin Correspondence Project*, Letter no. 8361). In the letter, he also suggested the name of a young artist, a friend of Butler's at Heatherley's, Arthur May. At the time, Darwin was looking for someone able to draw pictures for his imminent book *The Expression of the Emotions in Man and Animals* (1872).[42] Butler showed some of May's drawings to Darwin who really appreciated the skills of the young artist and decided to later include some of his work in the book.

Another letter worth mentioning was sent by Butler to Darwin in 1873 to talk about his new book on religion: *The Fair Haven* (a satirical piece of writing aimed at defending Christianity, which instead undermined its foundations). Butler sent along a copy of the volume to Darwin, who replied again with enthusiasm praising the quality of the new piece of work. *The Fair Haven* was so well received that on 1 April 1873, Darwin invited Butler to visit him again, saying 'if you could come here we should have been very glad to have seen you at luncheon or dinner',[43] probably to further discuss the writer's critical opinions about religion. Butler enthusiastically replied on 15 April 1873 and then visited Darwin and his family.[44]

In addition to Charles, Butler also became close to his son Francis. As reported by Henry Festing Jones' *Memoir*, the writer and Francis often met in London and were in the habit of dining together and going to concerts.[45] During those meetings, as Jones wrote, they discussed scientific topics, especially the ones close to Butler's interests, including evolution, design and Lamarckism, which would later become central to the publication of *Life and Habit*.

The publication of *Life and Habit* is particularly important here, because it created, for the first time, a distance between Butler and Charles Darwin. Concerning *Life and Habit*, there is a series of letters that deserve attention. On 24

September 1877, Butler sent to Charles Darwin, via his son, the manuscript of his first book on science alongside a long letter (*Darwin Correspondence Project*, Letter no. 11152). Butler's desire was to inform Charles Darwin about his idea of 'unconscious memory', and he hoped to receive comments and suggestions about his theory. Two months later, on 25 November 1877, Butler wrote another letter to Francis announcing the imminent publication of *Life and Habit*. In the letter, he offered to send two copies of his book, one for Francis and one for his father Charles. He wrote:

> I am going down home this week but I expect before I return my book will be out; it has been vexatiously delayed by printers, but should leave the binders on Thursday or Friday, and I have left the instruction that two copies shall be at once sent you — one of which if you think fit after reading it, you will perhaps be kind enough to give it to your father.[46]

Butler was obviously worried about the use of the work of Lamarck and Mivart in the development of his theory of memory and heredity. As had happened before with *Erewhon*, where reviewers had understood his satirical writing about evolution as a critique of Darwin's theory of natural selection, Butler was concerned about the possible misunderstandings that his ideas could generate among the Victorian professional and lay readerships. Additionally, this time Butler's book did not express full support for Darwin and his theory. Instead, as indicated by Butler himself, the aim was to complement Darwin's work with a return to the notion of inheritance. In *Life and Habit*, Butler stressed the necessity to create a link between natural selection and the work of Lamarck:

> With these additions (if they are additions) I cannot see that Lamarck's system is wrong. As for 'natural selection' frankly, it now seems to me a rope of sand as in any way accounting for the origin of species. Of course I am strengthened in my opinion by seeing that it [*Life and Habit*, CT] reduces to a common source the sterility of hybrids; the sterility of many wild animals under domestication; all variation (as being only a phase of sterility itself or rather the only alternative left to a creature under greatly changed conditions if the changes are not great enough to induce sterility); the phenomena of growth and metagenesis; the phenomena of old age; and a lot more which I see at present too uncertainly to venture to commit myself to paper concerning them.[47]

On 28 November 1877, Francis replied to Butler, concerned about his stern judgment on natural selection. In the letter, Francis especially referred to Butler's use of Mivart's work, defining it as follows:

> I think the falseness of Mivart's argument (Genesis of Sp. chapter ii.) is shown by applying it to man's selection which we see before our eyes at work. [...] I think if Mivart had been more of a naturalist instead of an anatomist he would not have dared to think that he could gauge natural selection's power of discrimination.[48]

Francis expressed some specific concerns about Butler's hypothesis of evolution — especially because of the latter's use of Mivart's teleological approach over that of Darwin. Sent one month later, on 28 December 1877, a second letter from Francis provides a more detailed explanation of those concerns. In the letter,

Darwin's son advanced a few criticisms about the conclusion of *Life and Habit*, signifying the inefficacy of Butler's argument and recommending, instead, Huxley's hypothesis of 'animal automatism'.[49] In particular, Francis suggested looking at the recent work of Huxley, where the scientist 'tried to show that consciousness was something superadded to nervous mechanism, like the striking of a clock is added to the ordinary going parts'.[50] According to Francis, although Butler's hypothesis regarding the analogy between memory and heredity was very well crafted, his psychological vision of evolution was not properly scientific or supported by enough solid evidence.

Butler replied on 29 December displaying some disappointment about the heavy-handed judgment of *Life and Habit*. He wrote: 'One line to thank you for yours of this morning, which I confess was rather a relief to me, as I was afraid you might have considered *Life and Habit* unpardonable.'[51] However, as observed by Henry Festing Jones, in '*Life and Habit* (December, 1877) it began to appear that Butler was dissatisfied with much in Charles Darwin's writings, but there was as yet no open breach between him and the Darwins'.[52] Butler's letter also contained a reference to the possibility of writing another book about evolution where he would explore the theories proposed and discussed before the *Origin of Species*.

Evolution, Old and New, The Quarrel and the Posthumous Reconciliation

This new book, entitled *Evolution, Old and New*, was published in London on 1 May 1879. Butler defined the content of *Evolution, Old and New* in a letter to his friend Ann Savage (5 November 1878) as follows:

> I am writing my Lamarck book [*Evolution, Old and New*] and am very full of it — but it is hard work, as it must be done very delicately, or I shall do old Darwin more good than harm. I shall never be happy till I have done it.[53]

Written within a year, *Evolution, Old and New* proposed a historical examination of the theories of evolution advanced before the publication of Darwin's *On the Origin of Species*. In particular, Butler's book introduced and summarized the work of three naturalists: Buffon, Erasmus Darwin, and Lamarck. For each of them, Butler's aim was to show how they anticipated some of the key aspects that made Darwin's work so original and innovative. If compared to Butler's previous writing on science, this is the first volume where the writer openly criticized Darwin's work and ideas.

Evolution, Old and New had two editions during Butler's lifetime. The second edition was published in 1882 and reissued with a new title page in 1890.[54] In his notebooks, Butler defined the scope of the volume as 'the tidying up the earlier history of the theory of evolution' and 'the exposure and discomfiture of Charles Darwin and Wallace and their followers'.[55] Partially a history of evolution and partially an example of popularization of science, Butler's *Evolution, Old and New* aimed at informing the Victorian readership about the complexity of the evolutionary debate and how various theories of evolution existed and were discussed long before the publication of Darwin's *Origin of Species*. Indeed, in the volume, Butler explained, referring to the general Victorian readership, that only 'few know that there are other great works upon descent with modification besides Mr. Darwin's. Not one person in ten thousand has any distinct idea of what Buffon, Dr. Darwin, and Lamarck propounded'.[56]

It is interesting to see how in Butler's *Evolution Old and New*, Darwin was described as nothing other than an illusionist who tried 'to throw dust in the eyes of those who would oppose the measure'.[57] The *Origin*, Butler claimed, presented several omissions, especially about the originality of the hypothesis of natural selection. Consequently, for Butler, the history of evolution had to be revisited, looking back at the work of William Paley, Buffon, Lamarck, Patrick Matthew (1790–1874), Étienne (1772–1844) and Isidore Geoffroy St. Hilaire (1805–1861), and Erasmus Darwin.

With *Evolution, Old and New* Butler embraced an ambitious task. On the one hand, he tried to jointly discuss two generations of naturalists, putting together different hypotheses and approaches. However, on the other hand, Butler saw *Evolution, Old and New* as both a chronology and a biographical account of the main protagonists of the evolutionary debate from Buffon to Charles Darwin. It follows that Butler tried to produce a historical examination of the evolutionary debate, while also attempting to provide further evidence in support of his idea of memory, heredity, and design.[58] In the preface to the second edition of *Evolution, Old and New*, Butler explains how the book relates to Darwin's work:

> I have insisted in each of my three books on Evolution upon the immensity of the service which Mr. Darwin rendered to that transcendently important theory. In 'Life and Habit,' I said: 'To the end of time, if the question be asked, "Who taught people to believe in Evolution?" the answer must be that it was Mr. Darwin.' This is true; and it is hard to see what palm of higher praise can be awarded to any philosopher.[59]

However, Butler also stresses that: 'No work can be judged intelligently unless not only the author's relations to his surroundings, but also the relation in which the work stands to the life and other works of the author, is understood and borne in mind.'[60] Indeed, *Evolution, Old and New* was conceived as an opportunity to partially rewrite the history of the discipline, putting an accent on its recent past. Therefore, to structure his new book, Butler decided to use a different approach if compared to *Life and Habit*. Instead of presenting and discussing his own theory, *Evolution, Old and New* was designed to be a more chronological examination of how evolution and, more in general, natural history developed from the mid-eighteenth century to his present day. For instance, in the chapters on Lamarck, the writer used a specific strategy to present his history of evolution to the readers: the section on the French naturalist starts with 'a memoir' which Butler translated directly from a biography of Lamarck, originally written in French, by M. Martins.[61] The biographical introduction was then followed by an examination of the reception of Lamarck's ideas among his contemporaries (including Darwin and Haeckel). It is here, I claim, that Butler's aim becomes clear. By looking at the reception of Lamarck's work, Butler's aim was to reveal how some of Lamarck's ideas were used, assimilated, and then forgotten by later naturalists including, among the many, Darwin and Huxley.

In this respect, *Evolution, Old and New* tried to persuade and remind his readers about the existence of a pre-Darwinian debate of evolution. However, this historical revival of Lamarck's ideas was also an opportunity to explain how evolution could be conceived from a teleological standpoint. So, the notion of design and the process of 'inheritance of acquired characteristics' obtained a new significance for Butler's theory of memory and heredity. This is confirmed in *Luck or Cunning?* where Butler writes:

> I therefore wrote 'Evolution Old and New,' with the object partly of backing up 'Life and Habit,' and showing the easy rider it admitted, partly to show how superior the old view of descent had been to Mr. Darwin's, and partly to reintroduce design into organism. I wrote 'Life and Habit' to show that our mental and bodily acquisitions were mainly stores of memory: I wrote 'Evolution Old and New' to add that the memory must be a mindful and designing memory.[62]

According to the quotation above, Butler's aim is to support his previous work but also to completely overturn the role of natural selection in evolution. Unfortunately, Butler's approach did not pay off. It does not take much imagination to see why Victorians agreed in dismissing Butler's book as the work of a Victorian who tormented the great 'Man of science' with slanderous accusations of dishonesty.[63] This general dismissal of Butler's work, is confirmed by the reviews of *Evolution, Old and New*. Although published by various journals and periodicals including *The Academy*, the *Examiner*, *The Field*, the *Daily Review*, *Nature*, *The Scotsman*, the *Daily News*, and *The Athenaeum* (among many others), they all saw Butler's work, at best, as a missed opportunity.

The first review of *Evolution, Old and New*, by Grant Allen, was published in *The Academy* on 17 May 1879. Here, Allen put into question Butler's expertise on the topic and even his ability to be a writer. Allen wrote:

> Mr. Butler comes forward, as it were, to proclaim himself a professional satirist, and a mystifier who will do his best to leave you utterly in the dark with regard to his system of juggling. Is he a teleological theologian making fun of evolution? Is he an evolutionist making fun of teleology? Is he a man of letters making fun of science? Or is he a master of pure irony making fun of all three, and of his audience as well? For our part we decline to commit ourselves, and prefer to observe, as Mr. Butler observes of Von Hartmann, that if his meaning is anything like what he says it is, we can only say that it has not been given us to form any definite conception whatever as to what that meaning may be.[64]

The review highlighted two critical aspects of Butler's *Evolution, Old and New* that made it difficult to be understood by both the professional and lay readership. First, Allen explained that because of both the structure and the argument of the book, it was difficult to clearly establish Butler's intentions. Second, Butler's volume did not provide any concrete evidence to support any of his claims (or anything recognizable as a professional demonstration) and therefore Butler's argument became unclear to any expert reader.

Similarly, a few days later *The Examiner* published another review, again, by Grant Allen.[65] As already stated in *The Academy*, Allen again dismissed Butler's work. However, this time he even decided to make fun of Butler's hypothesis:

> As to his (Mr. Butler's) main argument, it comes briefly to this: natural selection does not originate favourable varieties, it only passively permits them to exist; therefore it is the unknown cause which produced the variations, not the natural selection which spared them, that ought to count as the mainspring of evolution. That unknown cause Mr. Butler boldly declares to be the will of the organism itself. An intelligent ascidian wanted a pair of eyes, so set to work and made itself a pair, exactly as a man makes a microscope; a talented fish conceived the idea of walking on dry land, so it developed legs, turned its swim bladder into a pair of lungs, and became an amphibian; an æsthetic guinea-fowl admired bright colours, so it bought a paint-box, studied Mr. Whistler's ornamental designs, and, painting itself a gilded and ocellated tail, was thenceforth a peacock. But how about plants? Mr. Butler does not shirk even this difficulty. The theory must be maintained at all hazards.... This is the sort of mystical nonsense from which we had hoped Mr. Darwin had forever saved us.[66]

In this case, Allen focused more on Butler's idea of design. For the reviewer, it was not clear how *Evolution, Old and New* advanced an argument in favour of a teleological reading of evolution. Allen explained that the 'unknown cause' postulated by Butler to justify his evolutionary idea did not satisfy any scientific requirement and became, consequently, a 'mystical nonsense'.

There is another review, published in *The Saturday Review*, which can shed some light on the critics who moved against Butler's lack of concrete scientific evidence. Here, Butler's work was rejected because his argument was not supported by a strong and clear methodology. The review states:

> not professing to have any particular competence in biology, natural history, or the scientific study of evidence in any shape whatever, and, indeed, rather glorying in his freedom from any such superfluities, he undertakes to assure the overwhelming majority of men of science, and the educated public who have followed their lead, that, while they have done well to be converted to the doctrine of the evolution and transmutation of species, they have been converted on entirely wrong grounds.[67]

In partial contrast to the examples cited above, there is a review by Alfred Russel Wallace published in *Nature* (12 June 1879). Interestingly, this review approached Butler's book in a different way. Instead of focusing on the non-professional status of the author, Wallace discussed predominantly the content and defined *Evolution, Old and New* an 'interesting and useful' book.[68]

Wallace considered Butler's hypothesis as 'an important and even a necessary supplement to the theory advocated by Mr. Darwin'.[69] However, Wallace's review, it is fair to say, partially misrepresented Butler's intention; instead of seeing *Evolution, Old and New* as an attempt to rewrite the history of evolution proposing a historical justification in support of a reintroduction of design, Wallace considered the volume as mostly an opportunity for the Victorian readership to learn a bit more about the past of this fascinating discipline. Wallace wrote:

> It is, nevertheless, an interesting and useful book, inasmuch as it gives a pretty full account of the theories and opinions of several authors whose writing are almost unknown to the present generation of naturalists.[70]

The review, however, also highlighted the key point of Butler's work: the need to create a link between the past and the present while informing the general public. Butler's work on the evolutionary debate in France did not finish with the publication of *Evolution, Old and New*. The importance of the Lamarckian hypothesis and the role it played in criticizing Darwin's theory of natural selection and professional science became even more central in some of Butler's later works: *Unconscious Memory* and *Luck or Cunning?*.

It is now time to return to the quarrel. As discussed above, the publication of *Evolution, Old and New* represents for Butler's writing on science the first open criticism to Darwin's idea of evolution. However, Butler's severe opinion on Darwin's use of Lamarck's work and the negative reception of Butler's book did not have an impact on the two authors' relationship. Instead, the latter was compromised by a relatively simple mistake that is explained in a letter which Butler wrote to the editor of *Kosmos*:

> Dear Sir,
> Will you kindly refer me to the edition of *Kosmos* which contains the text of Dr. Krause's article on Dr. Erasmus Darwin, as translated by Mr. W. S. Dallas?
> I have before me the last February number of *Kosmos*, which appears by your preface to be the one from which Mr. Dallas has translated, but his translation contains long and important passages which are not in the February number of *Kosmos*, while many passages in the original are omitted in the translation.
> Among the passages introduced are the last six pages of the English article, which seem to condemn by anticipation the position I have taken as regards Erasmus Darwin in my book *Evolution Old and New*, and which I believe I was the first to take. The concluding, and therefore, perhaps, most prominent sentence of the translation you have given to the public stands thus: —
> 'Erasmus Darwin's system was in itself a most significant first step in the path of knowledge his grandson has opened up for us, but to wish to revive it at the present day, as has actually been seriously attempted, shows a weakness of thought and a mental anachronism which no one can envy.'
> The *Kosmos* which has been sent me from Germany contains no such passage.
> As you have stated in your preface that my book, *Evolution Old and New*, appeared subsequently to Dr. Krause's article, and as no intimation is given that the article has been altered and added to since its original appearance, while the accuracy of the translation, as though from the February number of *Kosmos* is, as you expressly say, guaranteed by Mr. Dallas's 'scientific reputation, together with his knowledge of German,' your readers will naturally suppose that all they read in the translation appeared in February last, and therefore before *Evolution Old and New* was written, and therefore independently of, and necessarily without reference to, that book.
> I do not doubt that this was actually the case, but have failed to obtain the edition which contains the passage above referred to, and several others which appear in the translation.

> I have a personal interest in this matter, and venture, therefore, to ask for the explanation, which I do not doubt you will readily give me. — Yours faithfully,
> S. BUTLER [71]

As is clear from the letter above, Butler was asking for some clarification regarding the omission of some reference to his work. In particular, Butler was concerned about discrepancies between an English (translation) and a German (original) version of the article. First, Butler noted that some sentences included in the German original had disappeared in the English translation. Second, and more problematic, the Victorian writer pointed out that specific additional and new passages, which had not been there in the original, had been added to the English translation. According to Butler, in particular, these new passages were aimed at undermining the claims he presented in *Evolution, Old and New*.

On 3 January 1880 Darwin responded to Butler:

> My dear Sir,
> Dr. Krause, soon after the appearance of his article in *Kosmos*, told me that he intended to publish it separately and to alter it considerably, and the altered MS. was sent to Mr. Dallas for translation. This is so common a practice that it never occurred to me to state that the article had been modified; but now I much regret that I did not do so. The original will soon appear in German, and I believe will be a much larger book than the English one; for, with Dr. Krause's consent, many long extracts from Miss Seward were omitted (as well as much other matter) from being in my opinion superfluous for the English reader. I believe that the omitted parts will appear as notes in the German edition. Should there be a reprint of the English Life, I will state that the original as it appeared in *Kosmos* was modified by Dr. Krause before it was translated. I may add that I had obtained Dr. Krause's consent for a translation, and had arranged with Mr. Dallas before your book was announced. I remember this because Mr. Dallas wrote to tell me of the advertisement. — I remain, Yours faithfully, C. DARWIN.[72]

Henry Festing Jones explained that 'Butler was not satisfied with this reply, and wrote to *The Athenaeum*, 31st January, 1880'.[73] Butler explained:

> At first I thought I ought to continue the correspondence privately with Mr. Darwin, and explain to him that his letter was insufficient, but on reflection I felt that little good was likely to come of a second letter, if what I had already written was not enough. I therefore wrote to the Athenæum and gave a condensed account of the facts contained in the last ten or a dozen pages. My letter appeared January 31, 1880.[74]

In the letter to the editor of *The Athenaeum*, Butler included a long list of what, in his opinion, had happened. This list included a summary of Krause's article in both German and English to show that in the English version eleven pages were missing.[75] Butler was, of course, convinced that this omission was not accidental. He believed that this was a deliberate attempt to undermine his work. This accusation is particularly clear in the conclusion of the letter where Butler stated:

> It is doubtless a common practice for writers to take an opportunity of revising their works, but it is not common when a covert condemnation of an

opponent has been interpolated into a revised edition, the revision of which has been concealed, to declare with every circumstance of distinctness that the condemnation was written prior to the book which might appear to have called it forth, and thus lead readers to suppose that it must be an unbiassed opinion.[76]

It is at this point that the situation became more interesting. Henry Festing Jones explains that when Darwin read Butler's letter, he realized that 'he had forgotten something' and that his 'instinct was to write to the *Athenæum*, and explain what had happened, but his intention was not carried into effect'.[77] Indeed, Darwin drafted two letters, that I will not include in their entirety but which can be consulted in Jones' pamphlet.[78] The first letter (dated 24 January 1880) is the longest one. Darwin seems to recognize a small problem in the translation. Darwin explains, in response to Butler's letter, that:

> He is mistaken in supposing that I was offended by this book, for I looked only at the part about the life of Erasmus Darwin; I did not even look at the part about evolution; for I had found in his former work that I could not make his views harmonize with what I knew. I was, indeed, told that this part contained some bitter sarcasms against me; but this determined me all the more not to read it.
>
> As Mr. Butler evidently does not believe my deliberate assertion that the omission of any statement that Dr. Krause had altered his article before sending it for translation, was unintentional or accidental I think that I shall be justified in declining to answer any future attack which Mr. Butler may make on me.[79]

The second letter (dated 1 February 1880) is far more interesting. Here, Darwin writes:

> In regard to the letter from Mr. Butler which appeared in your columns last week under the above heading, I wish to state that the omission of any mention of the alterations made by Dr. Krause in his article before it was re-published had no connection whatever with Mr. Butler. I find in the first proofs received from Messrs. Clowes the words: 'Dr. Krause had added largely to his essay as it appeared in *Kosmos*.' These words were afterwards accidentally omitted, and when I wrote privately to Mr. Butler I had forgotten that they had ever been written. (I could explain distinctly how the accident arose, but the explanation does not seem to me worth giving.) This omission, as I have already said, I much regret. It is a mere illusion on the part of Mr. Butler to suppose that it could make any difference to me whether or not the public knew that Dr. Krause's article had been added to or altered before being translated. The additions were made quite independently of any suggestion or wish on my part.[80]

As reported by Henry Festing Jones, Darwin was dissuaded by his family to send either of the letters to *The Athenaeum*.[81] Very unhappy with the situation, Darwin decided to ask Huxley's opinion. On 2 of February 1800, Darwin wrote:

> My dear Huxley, — I am going to ask you to [do] me a great kindness. Mr. Butler has attacked me bitterly, in fact, accusing me of lying, duplicity, and God knows what, because I unintentionally omitted to state that Krause had enlarged his *Kosmos* article before sending it for translation. I have written the enclosed letter [...] to the *Athenæum*, but Litchfield is strongly opposed to my making any answer, and I enclose his letter, if you can find time to read it. Of

the other members of my family, some are for and some against answering. I should rather like to show that I had intended to state that Krause had enlarged his article. On the other hand a clever and unscrupulous man like Mr. Butler would be sure to twist whatever I may say against me; and the longer the controversy lasts the more degrading it is to me. If my letter is printed, both the Litchfields want me to omit the two sentences now marked by pencil brackets, but I see no reason for the omission.[82]

As is clear from the above quotation, the quarrel was becoming more than a family dispute. On 3 February 1880, Huxley replied and advised Darwin to follow the advice of his family and to not send anything to *The Athenaeum*.[83] This was such a relief for Darwin, who, in his final letter on the matter, addressed to Huxley (4 February 1880), declared without hesitation:

Oh Lord what a relief your letter has been to me. I feel like a man condemned to be hung who has just got a reprieve. I saw in the future no end of trouble, but I feared that I was bound in honour to answer. If you were here I would show you exactly how the omission arose.[84]

In 1880 Butler published *Unconscious Memory*. In the volume he returned to the quarrel but only to stress once again his accusation. From this point onward, the quarrel moved from being a dispute over a missing reference to a proper open public argument about the intrinsic nature of evolution and heredity. James Paradis explains that Darwin's friends and allies felt the need to take part in the quarrel. Indeed, he writes: 'Butler's effort to mount a vigorous secular critique of natural selection in the Victorian periodical press was thus thoroughly demolished by an extensive network of Darwin's supporters, who effectively discredited him as a thinker and reasonable critic.'[85] From this moment onward, any attempt made by Butler to discuss Darwin's work or promote his own idea of evolution was dismissed as just a vague unprofessional attempt to debate a topic far too big for his own understanding.

This debate has been explored by David Gillott's recent book: *Samuel Butler against the Professionals* (2015). Gillott offers an overview of the problematic relationship between Butler and other professionals, tracing the history of Butler's life through the analysis of the evolution of his epistemological knowledge. Gillott uses the Darwin-Butler relationship and associated quarrel in order to show how Butler moved from believing in the professional objectivity of knowledge to considering professionals as individuals interested only in their personal careers.[86] By the end of his life, Butler was significantly outside any respectable professional community. The author was also not involved in any activities within any university or professional body. Butler was aware of the risk of working in an institution, considering it an obstruction of his freedom. In 1881, Butler wrote, in *Alps and Sanctuaries*, a very negative statement about university and the academic life:

Universities and academies are an obstacle to the finding of doors in later life; partly because they push their young men too fast through doorways that the universities have provided, and so discourage the habit of being on the look-out for others; and partly because they do not take pains enough to make sure that their doors are bona fide ones.[87]

The quarrel between Butler and Darwin was therefore more than a brief controversy. The quarrel directly influenced the reception of Butler's work and persona in Victorian England. A solution to the quarrel would arrive only a few years later when Francis Darwin published a revised edition of the *Life of Erasmus Darwin*. Henry Festing Jones clearly explains:

> Mr. Francis Darwin published a new edition of *Erasmus Darwin*, and fulfilled his father's promise to Butler by adding to the preface a third footnote: Mr. Darwin accidentally omitted to mention that Dr. Krause revised, and made certain alterations to, his Essay before it was translated.[88]

In 1901, Francis Darwin even publicly recognized the value of Butler's theory in a lecture he delivered at the Glasgow Meeting of the British Association 'On the Movements of Plants'.[89] Unfortunately, it was too late; Butler's writing on science — after having been demolished and ridiculed by professional Darwinists for so many years — was long forgotten, at least in Great Britain. The relationship between Butler and other professionals will be discussed in the next chapter. In particular, I will look at the place of Butler's in the nineteenth-century 'marketplace of science' and his remarks on George Romanes' theory of mental evolution.

Notes to Chapter 3

1. *Samuel Butler: Records and Memorials*, ed. by Richard A. Streatfeild (Cambridge: printed for private circulation, 1903) p. 9; also available in 'Samuel Butler', *The Athenaeum*, 3896 (28 June 1902), 819–20.
2. James Paradis, 'Butler after Butler: The Man of Letters as Outsider', in *Samuel Butler, Victorian against the Grain*, ed. by Paradis, pp. 343–70 (pp. 343–45).
3. Paradis, 'Butler after Butler', p. 345.
4. Paradis, 'Butler after Butler', p. 370; also included in Butler, *The Notebooks*, p. 368.
5. Raby, *Samuel Butler: A Biography*, p. 295.
6. Out of the ten obituaries that Richard A. Streatfeild, Butler's friend and testament executor, conserved for private circulation, one was from a Cambridge newspaper, *The Eagle*, one from New Zealand, *The Press*, and five articles were from Italy (Varallo, Trapani, and Rome). Streatfeild selected these specific papers because they were published in the three countries where the Victorian had lived and worked, but also because they — and in particular the Italian publications — offered an overview of how Butler's work was loved and discussed with interest outside Britain. See *Samuel Butler: Records and Memorials*, ed. by Streatfeild, pp. 1–3.
7. *Samuel Butler: Records and Memorials*, ed. by Streatfeild, p. 3.
8. *Samuel Butler: Records and Memorials*, ed. by Streatfeild, p. 4.
9. *Samuel Butler: Records and Memorials*, ed. by Streatfeild, p. 7.
10. *Samuel Butler: Records and Memorials*, ed. by Streatfeild, p. 7.
11. *Samuel Butler: Records and Memorials*, ed. by Streatfeild, p. 7.
12. Excluding Barlow's appendix (in *The Autobiography of Charles Darwin, 1809–1882*, ed. by Barlow, pp. 167–219), there are two main accounts of the quarrel: Basil Willey, *Darwin and Butler: Two Versions of Evolution. The Hibbert Lectures of 1959* (London: Chatto & Windus, 1960), and the more recent Paradis, 'The Butler-Darwin Biographical Controversy'.
13. Barlow, *The Autobiography of Charles Darwin, 1809–1882*, p. 170. See also Henry Festing Jones, *Charles Darwin and Samuel Butler: A Step Towards Reconciliation* (London: Fifield, 1911); Paradis, 'The Butler-Darwin Biographical Controversy'; Pauly, 'Samuel Butler and his Darwinian Critics', p. 161.
14. Barlow, *The Autobiography of Charles Darwin, 1809–1882*, p. 168.

15. Barlow, *The Autobiography of Charles Darwin, 1809–1882*, p. 172.
16. Barlow, *The Autobiography of Charles Darwin, 1809–1882*, p. 217.
17. Barlow, *The Autobiography of Charles Darwin, 1809–1882*, pp. 217–18.
18. Paradis, 'The Butler-Darwin Biographical Controversy', pp. 310–15.
19. Paradis, 'The Butler-Darwin Biographical Controversy', pp. 320–23.
20. Paradis, 'The Butler-Darwin Biographical Controversy', p. 324.
21. Paradis, 'The Butler-Darwin Biographical Controversy', pp. 324–25.
22. A number of letters that were exchanged between Butler and Darwin (1865–80) is held by the Cambridge University Library, Department of Manuscripts and University Archives, see NRA catalogue: NRA 11458 Darwin.
23. For further information regarding the relationship between Butler and Darwin prior to the quarrel refer to the two titles by Henry Festing Jones: *Samuel Butler: A Sketch* (London: Cape, 1913) and *Charles Darwin and Samuel Butler*.
24. Jones, *Samuel Butler, Author of Erewhon (1835–1902)*, I, 267–93.
25. Francis Darwin, *Life and Letters of Charles Darwin* (London: Murray, 1887), p. 168.
26. Charles Darwin to an (unnamed) editor, 24 March (1863?), in *Darwin Correspondence Project*, Letter no. 4058 <https://darwinproject.ac.uk/letter/?docId=letters/DCP-LETT-4058.xml> [accessed 6 January 2020]; all further references to this database will be given by letter number only.
27. The editor and journal to which Darwin sent this letter have not been identified; the letter was discovered in 1911, among the papers of John Malcolm Forbes Ludlow, a Christian socialist with an extensive acquaintance. See Butler, *A First Year in Canterbury Settlement*, pp. 149–55.
28. Haast was a German-born explorer and geologist. He travelled to New Zealand in 1858 to report on the prospects for German emigration. He was a friend of Samuel Butler, as reported in the record of his geological expedition (Julius Haast, *Geology of the Provinces of Canterbury and Westland* (Christchurch: printed at the *Times* Office, 1879), p. 4) and, later on, by Henry Festing Jones in his memoir (Jones, *Samuel Butler, Author of Erewhon (1835–1902)*, I, 70–86). Haast, also, founded the Philosophical Institute of Canterbury in 1862 and the Canterbury Museum in 1863. He was Professor of Geology at Canterbury College from 1876 to 1887 and Member of the senate of the University of New Zealand from 1879 to 1887. He was knighted in 1886 and became a Fellow of the Royal Society in 1867.
29. Charles Darwin to John (Julius) Haast, 18 July 1863, *Darwin Correspondence Project*, Letter no. 4245.
30. Emma Darwin to J. D. Hooker, 7 December 1863, *Darwin Correspondence Project*, Letter no. 4351. The squibs were published in *The Press* on 13 June 1863, p. 1, and 15 September 1863, p. 2. There are copies in the 'Scrapbook of Reviews' in the Darwin papers held at Cambridge University Library, Department of Manuscripts and University Archives, see DAR 226.1: 130 and DAR 226.1: 133–34, respectively.
31. Henry Festing Jones in his memoir referred to a long friendship between Butler and Haast, evidenced by many letters. See Jones, *Samuel Butler, Author of Erewhon (1835–1902)*, I, 70–86.
32. Haast, *Geology of the Provinces of Canterbury and Westland*, p. 4; also quoted in Jones, *Samuel Butler, Author of Erewhon (1835–1902)*, I, 88.
33. As indicated in a letter by Butler to Lilian Jones, dated 6 March 1902, Butler reported that in 1887, during the last visit of Haast to England, he gave all of his work on science to the geologist. See Jones, *Samuel Butler, Author of Erewhon (1835–1902)*, II, 386.
34. Charles Darwin to Samuel Butler, 30 September 1865, *Darwin Correspondence Project*, Letter no. 4902.
35. Charles Darwin to Samuel Butler, 1 October 1865, *Darwin Correspondence Project*, Letter no. 4904.
36. Charles Darwin to Julius Haast, 26 December 1865, *Darwin Correspondence Project*, Letter no. 4956.
37. See review of Samuel Butler, *Erewhon: or, Over the Range*, in *Athenaeum*, 2321 (20 April 1872), 492.
38. See review of Samuel Butler, *Erewhon: or, Over the Range*, in *Fortnightly Review*, 11.65 (May 1872), 609–10.

39. See review of Samuel Butler, *Erewhon Erewhon: or, Over the Range*, in *British Quarterly Review*, 56 (1872), 261–63.
40. Samuel Butler to Charles Darwin, 11 May 1872, *Darwin Correspondence Project*, Letter no. 8318.
41. Jones, *Samuel Butler, Author of Erewhon (1835–1902)*, I, 157.
42. Samuel Butler to Francis Darwin, [before 30 May 1872], *Darwin Correspondence Project*, Letter, no. 8305.
43. Samuel Butler to Charles Darwin, 1 April 1873, *Darwin Correspondence Project*, Letter no. 8835.
44. Samuel Butler to Charles Darwin, 15 April 1873, *Darwin Correspondence Project*, Letter no. 8859.
45. Jones, *Samuel Butler, Author of Erewhon (1835–1902)*, I, 256–57.
46. Samuel Butler to Francis Darwin, 25 November 1877, *Darwin Correspondence Project*, Letter no. 11152.
47. Samuel Butler to Francis Darwin, 25 November 1877, *Darwin Correspondence Project*, Letter no. 11152.
48. Letter from Francis Darwin to Samuel Butler, 28 November 1877. The letter is also reported in Jones, *Samuel Butler, Author of Erewhon (1835–1902)*, I, 261.
49. Jones, *Samuel Butler, Author of Erewhon (1835–1902)*, I, 261.
50. Jones, *Samuel Butler, Author of Erewhon (1835–1902)*, I, 263.
51. Jones, *Samuel Butler, Author of Erewhon (1835–1902)*, I, 264.
52. Barlow, *The Autobiography of Charles Darwin, 1809–1882*, p. 176.
53. Samuel Butler, *Letters between Samuel Butler and Miss E. M. A. Savage, 1871–1885* (London: Cape, 1935), pp. 195–96.
54. Of particular interest here is the preface of the second edition of *Evolution, Old and New*. There Butler wrote, referring to Darwin's death: 'Since the proof-sheets of the Appendix to this book left my hands, finally corrected, and too late for me to be able to recast the first of the two chapters that compose it, I hear, with the most profound regret, of the death of Mr. Charles Darwin. It being still possible for me to refer to this event in a preface, I hasten to say how much it grates upon me to appear to renew my attack upon Mr. Darwin under the present circumstances.' See Samuel Butler, *Evolution, Old and New, or, The Theories of Buffon, Dr. Erasmus Darwin and Lamarck, as Compared with that of Charles Darwin* (London: Fifield, 1911), p. vii.
55. Butler, *The Notebooks*, p. 375.
56. Butler, *Evolution, Old and New*, p. 61.
57. Butler, *Evolution, Old and New*, p. 358.
58. Significantly, in *Evolution, Old and New* Butler used the work of William Paley and of those who he defines as the 'first evolutionists' to justify his theory of evolution. Butler confirms this at the beginning of the book: 'Can we or can we not see signs in the structure of animals and plants, of something which carries with it the idea of contrivance so strongly that it is impossible for us to think of the structure, without at the same time thinking of contrivance, or design, in connection with it? It is my object in the present work to answer this question in the affirmative, and to lead my reader to agree with me, perhaps mainly, by following the history of that opinion which is now supposed to be fatal to a purposive view of animal and vegetable organs. I refer to the theory of evolution or descent with modification.' See Butler, *Evolution, Old and New*, pp. 1–2.
59. Butler, *Evolution, Old and New*, p. vii.
60. Butler, *Evolution, Old and New*, p. viii.
61. Butler, *Evolution, Old and New*, p. 235.
62. Butler, *Luck or Cunning?*, pp. 22–23.
63. See Raby, *Samuel Butler: A Biography*, pp. 161–78.
64. Grant Allen, review of Samuel Butler, *Evolution, Old and New*, in *Academy*, 15 (1879), 426.
65. In the published edition of his correspondence Butler added the following sentence as a postscript: 'I had sent her two reviews of *Evolution Old and New* — one in the *Academy* signed by Grant Allen. The other in the *Examiner* unsigned, but also (so D'Avigdor, the then editor, told me) by Grant Allen. Both reviews appeared on the same day, May 17th — and they led off the reviews. S.B.', see Butler, *Letters between Samuel Butler and Miss E. M. A. Savage*, p. 201.
66. Grant Allen, review of Samuel Butler, *Evolution, Old and New*, in *The Examiner* (17 May 1879), 646–47 (p. 646).

67. Review of Samuel Butler, *Evolution, Old and New*, in *Saturday Review* (31 May 1879), 682–84 (p. 682).
68. Alfred Russel Wallace, review of Samuel Butler, *Evolution, Old and New*, in *Nature*, 20, (1879), 141–44 (p. 141).
69. Wallace, review of Butler, *Evolution, Old and New*, p. 141.
70. Wallace, review of Butler, *Evolution, Old and New*, p. 142.
71. Barlow, *The Autobiography of Charles Darwin, 1809–1882*, pp. 178–79.
72. Barlow, *The Autobiography of Charles Darwin, 1809–1882*, pp. 179–80.
73. Barlow, *The Autobiography of Charles Darwin, 1809–1882*, p. 179.
74. Butler, *Unconscious Memory*, pp. 50–51.
75. Barlow, *The Autobiography of Charles Darwin, 1809–1882*, p. 181.
76. Barlow, *The Autobiography of Charles Darwin, 1809–1882*, p. 182.
77. Barlow, *The Autobiography of Charles Darwin, 1809–1882*, p. 182.
78. Barlow, *The Autobiography of Charles Darwin, 1809–1882*, pp. 182–86.
79. Barlow, *The Autobiography of Charles Darwin, 1809–1882*, p. 184.
80. Barlow, *The Autobiography of Charles Darwin, 1809–1882*, p. 185.
81. Barlow, *The Autobiography of Charles Darwin, 1809–1882*, p. 186.
82. Barlow, *The Autobiography of Charles Darwin, 1809–1882*, p. 186.
83. The letter is included in the *Darwin Correspondence Project*, Letter no. 12457.
84. Barlow, *The Autobiography of Charles Darwin, 1809–1882*, pp. 187–88.
85. Paradis, 'The Butler-Darwin Biographical Controversy', p. 325.
86. See David Gillott, *Samuel Butler against the Professionals: Rethinking Lamarckism 1860–1900* (London: Legenda, 2015), pp. 49–81.
87. Samuel Butler, *Alps and Sanctuaries of Piedmont and the Canton Ticino* (London: Bogue, 1882), p. 200.
88. Barlow, *The Autobiography of Charles Darwin, 1809–1882*, pp. 193–94.
89. A full account of the lecture was published in *Nature*, and included the following sentence: 'If we take the wide view of memory which has been set forth by Mr. S. Butler (*Life and Habit*, 1878) and by Professor Hering, we shall be forced to believe that plants, like all other living things, have a kind of memory.' See Barlow, *The Autobiography of Charles Darwin, 1809–1882*, p. 94.

CHAPTER 4

An Amateur among the Professionals

> Few know that there are other great works upon descent with modification besides Mr. Darwin's. Not one person in ten thousand has any distinct idea of what Buffon, Dr. Darwin, and Lamarck propounded. Their names have been discredited by the very authors who have been most indebted to them; there is hardly a writer on evolution who does not think it incumbent upon him to warn Lamarck off the ground which he at any rate made his own, and to cast a stone at what he will call the 'shallow speculations' or 'crude theories' or the 'well-known doctrine' of the foremost exponent of Buffon and Dr. Darwin.
>
> SAMUEL BUTLER, *Evolution, Old and New*[1]

The circulation of scientific ideas among both professionals and the general public was a key aspect of the Victorian scientific debate. In the form of books, essays, pamphlets, and public lectures, questions about science and its secrets fully penetrated the philosophical, cultural, and literary debates both in England and abroad. The periodical press was full of articles about the work of figures such as Charles Darwin, Alfred Russel Wallace, and Thomas Huxley. However, the work of certain individuals remained largely unknown in England even though they started to be successfully discussed in other countries. One such example was Butler's science of the mind. Butler's attempt to popularize and rediscuss Lamarckism, teleological design, and the substantial relationship between organic memory and heredity received little if any acknowledgement from the Victorian scientific community. It is, therefore, necessary to further explore the Victorian scientific debate starting from its popular components in order to fully understand why and how Butler's writings failed to impress both professionals and the general public alike.

As suggested by Lightman's '"The Voices of Nature": Popularising Victorian Science' (2000), although the simultaneous professional codification of science and popular writing worked on the same themes, popular scientific writing in the period is best understood as a product of the conflict of these two communities over a common subject.[2] Similarly, in her book *Science and Salvation* (2004) Aileen Fyfe explains that the word popular

> was not a description of the reception of a work (for example, a book was popular because everyone read it and liked it), but a statement about the intended audience envisaged by writers and publishers. [...] A 'popular'

work was one that was intended for 'the people,' which by the middle of the nineteenth century increasingly included the working classes.[3]

Indeed, Fyfe claims that we cannot understand ideas and attitudes towards science in the Victorian age by concentrating solely upon elite scientists and authors.[4] Therefore, the label 'popular' involves the presence of a mass audience for Victorian scientific writing and the acceptance of a new type of writer working outside any professional groups of scientists and authors.[5]

In order to understand the place and significance of Butler's popular writing on science among other Victorian professionals and amateurs we need to explore the differences between professional scientists and popularizers in terms of their usage of different languages (scientific and popular); places for research (public and private laboratories, museums, and universities); as well as audiences and reception.[6] Our context here is the Victorian marketplace of science as recently framed by the collection of essays *Science in the Marketplace* (2007).[7]

In the essay 'Publishing "Popular Science" in Early Nineteenth-Century Britain' (2007), Jonathan Topham suggests that 'there has never been a 'unified' or 'uncontested "popular science"'' whose boundaries are clearly demarcated.[8] Instead, in the nineteenth century, the word 'popular' began to be used to describe publications for those without specialist or expert knowledge and also to define publications intended for a mass audience. Consequently, the two different senses of the term 'popular' did not always coincide, and together with the emergence of 'science' as a specialist activity they reflected 'a growing sense of the disintegration of a unitary bourgeois public [...] and of the diversification of reading audiences'.[9] Therefore, 'popular science' did not contribute to the creation of new knowledge, and the epithet 'popular' involved something different from 'science'.[10]

Butler and the Marketplace of Science

The establishment of popular science in the nineteenth century was defined by two different, yet related, revolutions. The first of these was brought about by editors and the publishing industries. Publishers in the nineteenth century were the protagonists of a two-part industrial change in printing and selling books. The initial phase took place between 1830 and 1850 and was characterized by the introduction of the steam-driven press, case binding, and the fourdrinier machine, as well as the birth of reading and discussion circles.[11] The second phase started around 1855 and involved a significant change in the number of books published per year and the consequent increase in sales.[12] Two titles were especially relevant in terms of selling and public reception for science in general and evolution in particular: *On the Origin of Species* and the *Vestiges of the Natural History of Creation*. By the end of the century, Chamber's book sold 39,000 copies, and Darwin's a total of 56,000 copies.[13]

The second revolution was the creation and establishment of a proper platform for the discussion of scientific ideas. In order to understand the dimension of the Victorian marketplace, one needs to look at the exchange of ideas among scientists, novelists, and more general thinkers in both public and private venues. The new public dimension of science involved debates in museums, social circles, libraries, public squares, or, more generally, any place in which it was possible to converse about science and its cultural consequences.[14] Discussions were also conducted in Victorian periodicals where both professional scientists and lay writers debated their ideas about the main scientific themes of the time. As a consequence, the marketplace involved a variety of individuals who actively contributed to a partially unified debate. First, there were scientists, doctors and physiologists including, to just name a few, Huxley, Wallace, Darwin, and John Tyndall (1820–1893) who publicly championed the rise of the professionalization of science.[15] The creation of professional science played a key role in the geography, access to, and distribution of knowledge within the marketplace.[16] As we will discuss later on, this change is particularly important for Butler's own understanding of knowledge production and the reception of his ideas among specific circles.

Nonetheless, in the marketplace, evolution and science were also the domain of non-specialists. The latter included philosophers and writers such as Charles Kingsley, Charles Alexander Johns (1811–1874), Francis Orpen Morris (1810–1893), Thomas Willian Webb (1807–1885), George Henslow (1835–1925), and William Houghton (1828–1895) who promoted a reading of evolution close to their philosophical ideas.[17] Novelists, including George Eliot (1819–1880), William Morris (1834–1896), Charles Dickens (1812–1870), and later H. G. Wells (1866–1946) and Bram Stoker (1847–1912), explored evolution in their fictions. Finally, there were those like Butler whose writing on science tried to be appealing to both professionals and the general public but unfortunately did not succeed in either attempt. It is no surprise, therefore, to see Bernard Lightman's 2007 essay '"A Conspiracy of One": Butler, Natural Theology, and Victorian Popularization' acknowledge the English writer as one of the most controversial members of the Victorian marketplace.[18]

I partially disagree with the idea put forward by Lightman. Yes, Butler was a very controversial amateur scientist, but his work was not just aimed at popularizing science.[19] Butler always wanted to be an active protagonist of the debate. For example, when, in 1876, he started writing his first scientific book aimed at a general audience, *Life and Habit*, Butler was convinced that a more popular explanation of Darwin's work was necessary. In defining the scope of the volume Butler stated:

> It is plain, therefore, that my book cannot be intended for the perusal of scientific people; it is intended for the general public only, with whom I believe myself to be in harmony, as knowing neither much more nor much less than they do.[20]

Life and Habit combined analytical research with immense comparative work. Butler discussed and compared the works and ideas of British naturalists, philosophers, psychologists, and physiologists with the ones of those working on the continent. In contrast to the work of Victorian professionals, Butler's argument was not based on the result of an experiment or the direct observation of any natural phenomena.

Butler's approach did not fit with the new professional tradition that was becoming central to research. Paul White's recent book on Huxley traces the history of the establishment of scientific professionalism in the 1870s by stressing the need, advanced by 'Darwin's Bulldog', to distinguish the creation of scientific knowledge from its popularization.[21] As White explains, Huxley's aim was to create the Victorian 'man of science': a professional researcher working in a laboratory with proper duties, goals, and tools.[22] Huxley's approach inevitably contributed to creating a distance between professionals and popularizers that did not exist before.

As explained by Lightman, the Victorian popularizer was very often not a practitioner and their work was 'mainly focused on writing about nature'.[23] There is, therefore, a distinction to make between the Victorian professional scientist, who worked in a specific institution among peers, and the popularizer who often worked in isolation and whose main job was to entertain the masses by talking about science.[24] On the specific case of Butler, Lightman, in the above-mentioned essay '"A Conspiracy of One"' shows the importance of Butler as a popularizer but limits the relevance of his work to the general public, excluding the professional debate.[25] In particular, Lightman suggests that Butler's Lamarckism was used to exercise criticism over Darwin and other Darwinian professionals.[26] Lightman's argument is only partially correct. This is because Butler's science of the mind, although relying heavily on the work of others, advanced an original contribution to the evolutionary debate. In doing so, Butler became effectively one of the so-called forgotten pioneers of psychology.[27] Lightman is, however, correct in presenting Butler's work as a threat to the emergence of professionalism.

It is worth citing again the controversial statement Butler made about the production of scientific knowledge in *Life and Habit*: 'I say that the term "scientific" should be applied (only that they would not like it) to the nice sensible people who know what's what rather than to the discovering class.'[28] As we can see, for Butler, making science was not simply a question of conducting experiments or collecting specimens in remote locations. Instead, the production of new scientific knowledge was possible simply by knowing and reflecting upon the ideas of others. Butler's work as an amateur offers a different approach to evolution. Without the limits imposed by being a professional, Butler was able to create a dialogue between science, art, and philosophy.[29]

The whole of *Life and Habit* is written as a paradox to keep the reader curious about the real intention of the author. In introducing the volume, Butler presented himself as a 'man of the street' without any:

> wish to instruct, and not much to be instructed; my aim is simply to entertain and interest the numerous class of people who, like myself, know nothing of science, but who enjoy speculating and reflecting (not too deeply) upon the phenomena around them.[30]

Butler's self-deprecatory denial of any professionalism in his work is only instrumental to his thought-provoking style of writing. Indeed, towards the end of the volume, Butler claimed a proper scientific value to his theory by comparing its quality with the work of other Victorian men of science. This made his writing difficult to categorize and even more difficult to be accepted by professional science.

Unconscious Memory and *Luck or Cunning?* present a similar approach. In *Unconscious Memory*, Butler combined writing on science and translations of foreign scientific papers with his personal criticism of Victorian professionalism. In *Luck or Cunning?*, Butler also pointed out his status as a 'non-man of science':

> Nevertheless I again freely grant that I am not a man of science. I have never said I was. I was educated for the Church. I was once inside the Linnean Society's rooms, but have no present wish to go there again.[31]

However, in the same volume, Butler also justified the scientific validity of his theory by stating that a: 'complaint against me is to the effect that I have made no original experiments, but have taken all my facts at second hand. [...] If the facts are sound, how can it matter whether A or B collected them?'[32] Butler's writing on science is, therefore, an example of speculative evolution, which mixes science with philosophy whilst often crossing the boundaries between the two.[33]

In this way, Butler was able to provide to his readers a comprehensive 'behind the scenes' account of the evolutionary debate, thus enabling a better understanding of his own hypothesis of evolution. Butler's writing on science tried to tell a new history to its readers by looking back to the past in order to understand the roots of more recent discoveries. In describing this approach, the author wrote in *Life and Habit*:

> I have therefore allowed myself a loose rein, to run on with whatever came uppermost, without regard to whether it was new or old; feeling sure that if true, it must be very old or it never could have occurred to one so little versed in science as myself; and knowing that it is sometimes pleasanter to meet the old under slightly changed conditions, than to go through the formalities and uncertainties of making new acquaintance.[34]

Butler's aim was to popularize science to his contemporaries but, also, to show them the mysteries and errors of current and past ideas of evolution. In order to achieve this, the Victorian writer used two methods: he translated the work of foreign physiologists, naturalists, and philosophers into English but also actively used those works to support his speculative ideas concerning evolution and the mind.

Writing about Science

In reading Butler's books on science, one of the first things that a reader notices immediately is the extensive use of quotations. In chapter two, I discussed Butler's use of different genres and the complexity of his writing. My preliminary conclusion presented Butler as a polymath who worked in various fields and genres. However, in the examination of the science of *Erewhon*, little attention was paid to the role played by translations of foreign texts in the development of Butler's own understanding of evolution.

In order to explore the importance of Butler's translation work, one should start with his upbringing. One of the key conclusions of Henry Festing Jones' *Memoir* concerning Butler's scientific work was the assumption that Butler's knowledge of continental languages helped him on several occasions in developing his idea of evolution. Butler was indeed self-trained in modern languages such as Italian, French, and German. With respect to science, Butler translated predominately the works of German and French naturalists and philosophers. For him, the scientific works produced in these two countries offered the opportunity to discuss evolution from a different perspective.[35]

In exploring Butler's work of translation, I rely on the definition provided by Jean Delisle and Judith Woodsworth in their historical work on translation as a practice. In *Translators through History* (1995), they define Victorian translators as a group of individuals who worked '[b]eyond the prerogatives of patrons, clients and editors, beyond the materiality of texts, beyond the cost of their labour, translators cross and blur the lines between foreign cultural values and those of their own society'.[36] This definition seems to overlap perfectly with what Butler did in all of his scientific books. He always tried to implement his own style even when translating and discussing the work of others. As evidence, let us take a look at how Butler, between 1878 and 1890, translated the work of German evolutionists and philosophers into English. Butler did not study German during his childhood as he did with French and Italian. He learned German mostly out of necessity. As reported in *Unconscious Memory*, when *Evolution, Old and New* was published, Butler started learning German in order to communicate with the editor of *Kosmos*. In *Unconscious Memory*, Butler included and commented on a specific passage referring to his work of translation:

> At this time I knew not one word of German. On the same day, therefore, that I sent for *Kosmos* I began to acquire that language, and in the fortnight before *Kosmos* came had got far enough forward for all practical purposes — that is to say, with the help of a translation and a dictionary, I could see whether or no [sic] a German passage was the same as what purported to be its translation.[37]

However, although Butler was not a specialist in translating German, he did his best to produce accurate work. In introducing the translation of Hering's lecture in *Unconscious Memory*, Butler stated:

> I will now lay before the reader a translation of Professor Hering's own words. I have had it carefully revised throughout by a gentleman whose native language is German, but who has resided in England for many years past.[38]

Once again, Butler approached the work of translating Hering's lecture in his own way. Butler's approach created two problems. First, instead of employing a professional translator, Butler decided that it was good enough for him to translate Hering's lecture. Although the translation was pretty accurate, it was not meant to be perfect. Second, the decision to include this important and necessary translation in his most polemical book contributed to hiding the value of Hering's work. Indeed, as we will soon see, *Unconscious Memory* was overlooked by both the professional and lay readership with the result that the significance of Hering's lecture was not taken into serious consideration.

There is another aspect of Butler's popular writing on science that needs attention: his engagement with the work of his peers. So far, I have predominately discussed Butler's opinion about Darwin and how it changed over time. The discussion of the quarrel, in particular, has shown how the complex relationship between Butler and the author of the *Origin of Species* was fundamental in the development of his ideas about evolution and heredity. However, in his writings on science, Butler also engaged with several other Victorian evolutionists. As mentioned previously, the use of the theories and claims of professionals was an important component of Butler's own approach to the production of new scientific knowledge. Butler did not work in any professional institute, did not collect any evidence or conduct any experiment to support his own theory. This attitude was not received well by the growing professional community. In addition, the quarrel he had with Darwin contributed to isolating his work and ideas. Therefore, a critical engagement with the work of other professionals was for the Victorian writer an opportunity to show how his role within the scientific debate, although undermined by his status as an amateur, still had a value.

So, in order to demonstrate to the Victorian scientific community the importance and significance of his ideas about memory, heredity, and Lamarckism, Butler decided to reveal how some aspects of his theory were an integral part of the work of other, well-known figures. Butler selected the work of three of his contemporaries: St. George Jackson Mivart, Herbert Spencer, and George Romanes. The author believed that by showing the similarity between his work and that of some well-known evolutionists, he would be able to justify the validity of his scientific claims about heredity and Lamarckism. The case of Romanes, in particular, is instrumental in exploring how Butler's work was unsuccessful due to his non-professional status and despite the fact that he proposed an idea of evolution not that different from the one suggested by Darwin's pupil.

Butler and St. George Jackson Mivart: From the *Origin of Species* to *Genesis of Species*

In 1871, St. George Jackson Mivart published *On the Genesis of Species*, a book that soon became the critic of natural selection which Charles Darwin took most seriously. In his work, Mivart tried to reconcile Darwin's theory of natural selection with the metaphysical interpretations of nature as used by the Catholic Church.[39] He became famous, like Butler, for being at first an ardent believer in natural selection, only to later become one of its fiercest critics. This shift is clear from Mivart's own words in 'Evolution and its Consequences', an article published in *The Contemporary Review* in January 1872.[40] Mivart explained:

> My 'Genesis of Species' was written with two main objects: My first object was to show that the Darwinian theory is untenable, and that natural selection is not the origin of species. This was and is my conviction purely as a man of science, and I maintain it upon scientific grounds only. My second object was to demonstrate that nothing even in Mr. Darwin's theory, as then put forth, and *a fortiori* in evolution generally, was necessarily antagonistic to Christianity.[41]

Mivart's attempt to reconcile evolution with the Christian understanding of the world unfortunately led to condemnation of his work by both parties.[42] Although Mivart's intention to discuss evolution in relation to questions concerning religion was clear from the introduction of *On the Genesis of Species*, his work was not received as expected.[43] Mivart wrote:

> Remarkable is the rapidity with which an interest in the question of specific origination has spread. But a few years ago it scarcely occupied the minds of any but naturalists. Then the crude theory put forth by Lamarck, and by his English interpreter, the author of the 'Vestiges of Creation,' had rather discredited than helped on a belief in organic evolution a belief — that is, in new kinds being produced from older ones by the ordinary and constant operation of natural laws. Now, however, this notion is widely diffused. Indeed, there are few drawing-rooms where it is not the subject of occasional discussion, and artisans and schoolboys have their views as to the permanence of organic forms. Moreover, the reception of this doctrine tends actually, though by no means necessarily, to be accompanied by certain beliefs with regard to quite distinct and very momentous subject-matter. So that the question of the 'Genesis of Species' is not only one of great interest, but also of much consequence.[44]

Indeed, Mivart, in his *Genesis of Species*, explained that Darwin's work on evolution presented some scientific difficulties which cannot be ignored by any naturalist.[45] These scientific difficulties were described as several methodological errors and examined by the Victorian biologist. Mivart tried to maintain:

> the position that 'Natural Selection' acts, and indeed must act; but that still, in order to account for the production of known kinds of animals and plants, it requires to be supplemented by the action of some other natural law or laws as yet undiscovered.[46]

It thus follows 'that the consequences which have been drawn from Evolution, whether exclusively Darwinian or not, to the prejudice of religion, by no means follow from it, and are in fact illegitimate'.[47]

For Mivart, a secular form of teleology was the constructive answer to the questions left open by Darwinian evolution. In the *Genesis of Species*, he clearly pointed out the necessity to find a natural law to explain the evolutionary process and not another religious prejudice. Partially making criticism of Darwin's natural selection and partially accepting it, Mivart was in the difficult position of being in the middle of a complex debate, as asserted by Thomas Huxley's article 'Mr Darwin's Critics' in the *Contemporary Review* (1871). Huxley's text, forty-two pages long, while defending Darwinian evolution, also discussed the work of Mivart and Wallace explaining the relevance of their positions. The article highlights two key aspects of Mivart's work compared to that of Wallace. First, as Huxley explained:

> Thus Mr. Mivart is less of a Darwinian than Mr. Wallace, for he has less faith in the power of natural selection. But he is more of an evolutionist than Mr. Wallace, because Mr. Wallace thinks it necessary to call in an intelligent agent — a sort of supernatural Sir John Sebright — to produce even the animal frame of man; while Mr. Mivart requires no Divine assistance till he comes to man's soul.[48]

Huxley defined Mivart's position as both controversial and contradictory. This was particularly the case with regard to Mivart's use of an 'intelligent agent'. Nonetheless, Huxley's article also shed some light on the intrinsically religious nature of Mivart's argument. Huxley wrote:

> I may assume, then, that the *Quarterly Reviewer* and Mr. Mivart admit that there is no necessary opposition between 'evolution, whether exclusively Darwinian or not,' and religion. But then, what do they mean by this last much-abused term? On this point the Quarterly Reviewer is silent. Mr. Mivart, on the contrary, is perfectly explicit, and the whole tenor of his remarks leaves no doubt that by 'religion' he means theology; and by theology, that particular variety of the great Proteus, which is expounded by the doctors of the Roman Catholic Church, and held by the members of that religious community to be the sole form of absolute truth and of saving faith.[49]

Assuming that the aim of Huxley was to critically assess the arguments proposed by the 'critics' of Darwinism, his analysis can help us to understand Mivart's conception of design. Starting from the examination of the metaphysical and ontological work of the Spanish theologian and philosopher Francisco Suarez (1548–1617) and moving on to how he was discussed by Mivart in the *Genesis*, Huxley tried to establish the origin and development of Mivart's idea. Here, 'Darwin's bulldog's' aim was to demonstrate how using a religious approach to discuss evolution, which should be treated uniquely as a scientific matter, creates several problems and incongruences because of the metaphysical difference between the minds of men and lower animals.

In the conclusion of the article, Huxley tried to further assess that:

> The *Quarterly Reviewer* and Mr. Mivart base their objections to the evolution of the mental faculties of man from those of some lower animal form upon what they maintain to be a difference in kind between the mental and moral faculties of men and brutes; and I have endeavoured to show, by exposing the utter unsoundness of their philosophical basis, that these objections are devoid of importance.[50]

Although, as stated by Huxley, the criticism made by Mivart and 'the *Quarterly Reviewer*' were — in his opinion — 'devoid of importance', for Darwin the situation was different. In the 1872 edition of the *Origin*, while giving answers to a number of criticisms and objections moved against his idea of natural selection, Darwin stated:

> First, why, if species have descended from other species by fine gradations, do we not everywhere see innumerable transitional forms? Why is not all nature in confusion, instead of the species being, as we see them, well defined? Secondly, is it possible that an animal having, for instance, the structure and habits of a bat, could have been formed by the modification of some other animal with widely-different habits and structure? Can we believe that natural selection could produce, on the one hand, an organ of trifling importance, such as the tail of a giraffe, which serves as a fly-flapper, and, on the other hand, an organ so wonderful as the eye? Thirdly, can instincts be acquired and modified through natural selection? What shall we say to the instinct which leads the bee to make cells, and which has practically anticipated the discoveries of profound mathematicians? Fourthly, how can we account for species, when crossed, being sterile and producing sterile offspring, whereas, when varieties are crossed, their fertility is unimpaired?[51]

In the sixth edition of *The Origin of Species*, Darwin paid particular attention to the main detractors of his hypothesis of evolution. The four points explained at the beginning of chapter six of the *Origin* developed into a large discussion regarding the position of the main detractors of his theory of natural selection. On Mivart, Darwin stated:

> A distinguished zoologist, Mr. St. George Mivart, has recently collected all the objections which have ever been advanced by myself and others against the theory of natural selection, as propounded by Mr. Wallace and myself, and has illustrated them with admirable art and force.[52]

Darwin took the objections proposed by Mivart so seriously that, at the beginning of the chapter dedicated to the criticism against his theory, he wrote:

> All Mr. Mivart's objections will be, or have been, considered in the present volume. The one new point which appears to have struck many readers is, 'that natural selection is incompetent to account for the incipient stages of useful structures.' This subject is intimately connected with that of the gradation of characters, often accompanied by a change of function, — for instance, the conversion of a swim-bladder into lungs, — points which were discussed in the last chapter under two headings. Nevertheless, I will here consider in some detail several of the cases advanced by Mr. Mivart, selecting those which are the most illustrative, as want of space prevents me from considering all.[53]

Darwin's response to Mivart developed into a thirty-page long analysis of some of the key criticisms advanced by the biologist. These included discussions about the failure of natural selection to explain the incipient stages of various anatomical organ structures and the inability of natural selection to explicate cases of parallel evolution. Darwin was also particularly concerned with Mivart's idea of evolution as a matter of design. In the very conclusion of chapter six, Darwin explained:

> He who believes that some ancient form was transformed suddenly through an internal force or tendency into, for instance, one furnished with wings, will be almost compelled to assume, in opposition to all analogy, that many individuals varied simultaneously. It cannot be denied that such abrupt and great changes of structure are widely different from those which most species apparently have undergone. He will further be compelled to believe that many structures beautifully adapted to all the other parts of the same creature and to the surrounding conditions, have been suddenly produced; and of such complex and wonderful co-adaptations, he will not be able to assign a shadow of an explanation. He will be forced to admit that these great and sudden transformations have left no trace of their action on the embryo. To admit all this is, as it seems to me, to enter into the realms of miracle, and to leave those of Science.[54]

It is fair to say that Mivart's work represented one of the sharpest and most discussed criticisms of Darwin's hypothesis of natural selection in the 1870s. This made Mivart's work immediately appealing to someone like Butler, who was looking for a thinker who partially shared some of his ideas about evolution. Alongside similar ideas concerning teleology and design in evolution, Butler and Mivart also experienced a similar destiny: they both rejected Darwin's theory of natural selection and went through a series of quarrels with loyal Darwinists, which consequently isolated them from the professional scientific community.[55]

As discussed, in *On the Genesis of Species,* Mivart recognizes that the idea of evolution presented by 'Mr. Darwin is perhaps the most interesting theory, in relation to natural science, which has been promulgated during the present century'.[56] However, Mivart also adds that the 'special Darwinian hypothesis [...] is beset with certain scientific difficulties, which must by no means be ignored, and some of which, I venture to think, are absolutely insuperable'.[57]

In order to solve these problems, Mivart's aim was to combine Darwin's theory with the doctrine of the Catholic Church or, more in general, religious ideas.[58] Mivart presented a theory of evolution based on the idea of design, which still made strong reference to the work of Lamarck. However, Mivart also examined and engaged with the works of several theologians and philosophers, such as Augustine of Hippo, Thomas Aquinas, and Francisco Suarez, to explore how their ideas about the world could be used to provide a better explanation of evolution and its mechanism. This is particularly relevant in the final chapter of the *Genesis of Species*, 'Theology and Evolution', where Mivart compared these old theories with the work of Darwin, Spencer, Wallace, and Huxley to show their ideas' respective strengths and weaknesses. This approach was particularly welcomed by Butler who used a very similar method in his writings on science.

In 1878, with the publication of *Life and Habit,* Butler discussed the work of Mivart for the first time but predominately in relation to Darwin and various questions concerning teleology and design. According to Butler, Mivart's work represented the best (contemporary) way for understanding the necessity of rethinking evolution in relation to design.[59] In order to explain how Mivart criticized Darwin's work, Butler presented to his reader a long comparison between *The Origin of Species* and *The Genesis of Species*. For the Victorian writer, one of the key differences between the two works was exemplified by their different understandings of the notion of biological variation. In *Life and Habit*, Butler was very careful in explaining that

> when we bear in mind that the variations, being supposed by Mr. Darwin to be indefinite, or devoid of aim, will appear in every direction, we cannot forget what Mr. Mivart insists upon, namely, that the chances of many favourable variations being counteracted by other unfavourable ones in the same creature are not inconsiderable. Nor, again, is it likely that the favourable variation would make its mark upon the race, and escape being absorbed in the course of a few generations, unless — as Mr. Mivart elsewhere points out, in a passage to which I shall call the reader's attention presently — a larger number of similarly varying creatures made their appearance at the same time than there seems sufficient reason to anticipate, if the variations can be called fortuitous.[60]

According to Butler, then, Darwin's theory of natural selection can work only if based on a system of random variations. Mivart, instead, preferred to consider variations as a part of process linked to 'a power in nature which would preserve and accumulate further beneficial resemblance'.[61] This power in nature becomes fundamental for justifying the process of evolution. Butler explains:

> He [Mivart] thinks — and I believe the reader will agree with him — that this process is too slow and too risky. What he wants to know is, how the insect came even rudely to resemble the object, and how, if its variations are indefinite, we are ever to get into such a condition as to be able to report progress, owing to the constant liability of the creature which has varied favourably, to play the part of Penelope and undo its work, by varying in some one of the infinite number of other directions which are open to it — all of which, except this one, tend to destroy the resemblance, and yet may be in some other respect even more advantageous to the creature, and so tend to its preservation.[62]

For Butler, the system proposed by Mivart was decisive in demonstrating the fallacies of Darwin's natural selection. However, Mivart's ideas were also instrumental in supporting his own Lamarckian conclusion that evolution was nothing else than a form of guided memory, which could be repeated over and over again and from generation to generation.

This point is confirmed in the final chapter of *Life and Habit*, where Butler stated:

> Evolution entirely unaided by inherent intelligence must be a very slow, if not quite inconceivable, process. Evolution helped by intelligence would still be slow, but not so desperately slow. One can conceive that there has been sufficient time for the second, but one cannot conceive it for the first.[63]

Although in *Life and Habit*, as discussed above, Butler takes into high consideration Mivart's opinion about Darwin and natural selection, he also disagrees with some of his claims about the role and significance of ethics and its relationship with science. These disagreements are particularly clear when Butler discusses Mivart's ethical position. Butler wrote:

> When Mr. Mivart deals with evolution and ethics, I am afraid that I differ from him even more widely than I have done from Mr. Darwin. He writes ('Genesis of Species,' p. 234): 'That "natural selection" could not have produced from the sensations of pleasure and pain experienced by brutes a higher degree of morality than was useful; therefore it could have produced any amount of "beneficial habits", but not abhorrence of certain acts as impure and sinful.'[64]

Butler seems to agree only partially with the author of the *Genesis of Species*. While Butler supports Mivart's theory of evolution, he strongly disagrees with the more philosophical aspects of the latter's approach. Butler's opinion on Mivart is probably due to two different, yet related, factors. First, Butler, in line with his previous writing about Christianity and religion, was still rejecting any link between the two. Consequently, although he recognized the merit of Mivart's scientific ideas, Butler was not willing, in his first book on science, to create a bridge between evolution and Christianity. Second, it is worth remembering that the aim of *Life and Habit* was to complement Darwin's work with a theory encompassing the notion of design and not to fully reject natural selection.

In *Unconscious Memory*, Butler cited and discussed Mivart's work, but only briefly. The reference to Mivart was, this time, not about the content of his writings but rather its significance for the late nineteenth-century evolutionary debate. Here, Butler explains the importance that the *Genesis of Species* had for the development of his idea and growing criticism of Darwin's work:

> When I had finished the 'Genesis of Species,' I felt that something was certainly wanted which should give a definite aim to the variations whose accumulation was to amount ultimately to specific and generic differences, and that without this there could have been no progress in organic development. I got the latest edition of the 'Origin of Species' in order to see how Mr. Darwin met Professor Mivart, and found his answers in many respects unsatisfactory.[65]

In *Unconscious Memory*, Butler's position is, of course, different from that of *Life and Habit*. Butler was not trying to complement Darwin's theory anymore, and therefore Mivart became one of few contemporary scientists that Butler considered worth reading:

> I thought it right that people should have a chance of knowing more about the earlier writers on evolution than they were likely to hear from any of our leading scientists (no matter how many lectures they may give on the coming of age of the 'Origin of Species') except Professor Mivart.[66]

In particular, for Butler, Mivart's book was important because:

> A book pointing the difference between teleological and non-teleological views of evolution seemed likely to be useful, and would afford me the opportunity I wanted for giving a *résumé* of the views of each one of the three chief founders of the theory, and of contrasting them with those of Mr. Charles Darwin, as well as for calling attention to Professor Hering's lecture.[67]

In *Luck or Cunning?*, references to the work of Mivart are again limited. However, in his last book on science, Butler made a clear statement about the influence that Mivart had on the writing of *Life and Habit*. In particular, Butler explained how Mivart's work proved to be extremely useful in understanding the difference between natural selection and the idea of *transformisme* advanced by Jean-Baptiste Lamarck in 1809 and then used by other pre-Darwinian evolutionists including Étienne Geoffroy Saint-Hilaire, Robert Grant (1793–1874), and Robert Chambers. Butler wrote: 'Before I had finished writing this book I fell in with Professor Mivart's "Genesis of Species," and for the first time understood the distinction between the Lamarckian and Charles-Darwinian systems of evolution.'[68]

In summary, Mivart with his *Genesis of Species* became for Butler a precious ally to question Darwin's theory of natural selection. However, the book also proved to be a good source for a discussion concerning the role of Lamarckism and design in looking for an alternative to the rise of Darwinism. Butler used the research done by Mivart in support of his own hypothesis of memory as a form of heredity. In concluding this section, I would like to cite *Luck or Cunning?* once more, where Butler clearly explained why the author of the *Genesis of Species* was so significant in the development of his scepticism about Darwin's materialism:

> Mivart was, as I have said, among the first to awaken us to Mr. Darwin's denial of design, and to the absurdity involved therein. He well showed how incredible Mr Darwin's system was found to be, as soon as it was fully realised, but there he rather left us. He seemed to say that we must have our descent and our design too, but he did not show how we were to manage this with rudimentary organs still staring us in the face.[69]

In other words, Butler praised the importance of Mivart's work for helping some Victorians to realize that there was more to evolution than Darwin's work.

Herbert Spencer: between Lamarckism and Mental Evolution

The second key influential evolutionist used by Butler to justify his own work was Herbert Spencer. Butler largely cited Spencer, especially in relation to his engagement with Lamarck's ideas and pioneering work on psychology. The relevance of Spencer's ideas to Butler's thinking is clearly explained in *Luck or Cunning?*, where Butler stated: 'I can, however, have no hesitation in saying that if I had known the "Principles of Psychology" earlier, as well as I know the work now, I should have used it largely.'[70] In his final book on science, Butler recognized the importance of Spencer not only for the development of his own theory but also for the development of psychology as a whole.

Trained as a philosopher, Spencer advanced an all-embracing interpretation of evolution as the progressive development of the physical world, biological organisms, the human mind, culture, and societies. He considered evolution as a fully comprehensive way for understanding the whole world. In addition to all of this, Spencer was also one of the most influential figures in the social and cultural debates of the late nineteenth century and contributed to the development of new disciplines such as sociology, anthropology, and psychology. This made an engagement with his ideas central and necessary for anyone willing to look at evolution from a philosophical, sociological, or psychological perspective.

From the beginning of his career, Spencer developed an interest in the debate concerning evolution. In 1852, seven years before Darwin's *Origin of Species*, Spencer published an essay entitled 'The Developmental Hypothesis', in which he presented his own take on the debate. Although members of the scientific community found Spencer's idea interesting from a theoretical standpoint, they did not take it into serious consideration as it lacked an effective mechanism for explaining the changes and variations occurring through the evolutionary process. Historian Ruth Burton explains that although Spencer was a 'professional amateur' he was not a professional scientist in the sense that term acquired in the 1870s but still so relevant to the debate to be a member of the famous X-Club. This membership gave him the chance to be included in of the most important scientific circles of the period.[71]

In 1857, Spencer published in John Chapman's *Westminster Review* an essay entitled 'Progress: Its Law and Cause', which would later become the basis of the *First Principles of a New System of Philosophy* (1862).[72] In the essay, the philosopher advanced a reading of evolution which combined insights from Samuel Taylor Coleridge's (1772–1834) essay 'The Theory of Life' (1818) and Karl Ernst Ritter von Baer's (1792–1876) laws of embryological development.[73] Spencer suggested that all structures in the universe, from a cell to a galaxy, develop from a simple, undifferentiated, homogeneity to a complex, differentiated, heterogeneity, while being part of a process of greater integration of the differentiated parts (a sort of blueprint). Spencer's hypothesis was based on a universal law, which could be applied to the greater system as much as to detailed and small biological organisms.

Significantly, in the article, the key mechanism of species transformation that Spencer recognized is a Lamarckian process of inheritance, based on the system of the use and disuse of various organs and body parts, whereby the resulting changes are transmitted to future generations. Spencer argued that the Lamarckian approach was necessary to explain 'higher' evolution, especially in relation to the social development of humanity.[74] In contrast to Darwin, Spencer held the viewpoint that evolution has a direction and an endpoint and involves a final state of equilibrium. As explained by Mike Hawkins' *Social Darwinism in European and American Thought, 1860–1945* (1997), Spencer insisted on professing a strong role for the idea of equilibrium as a basic part of the process of evolution, but he also recognized the importance of natural selection.[75] Indeed, as reported in his autobiography, Spencer acknowledged the role and significance of Darwin in developing a new and convincing theory of evolution. He wrote:

> At that time I ascribed all modifications to direct adaptation to changing conditions; and was unconscious that in the absence of that indirect adaptation effected by the natural selection of favourable variations, the explanation left the larger part of the facts unaccounted for.[76]

In order to understand why Butler used Spencer's work to justify his own theory, it is necessary to explore the similarities and differences in their understanding and use of Lamarck's theory of evolution. This comparison will help to explain how and why the English philosopher directly influenced Butler in the development of his hypothesis of 'unconscious memory'. However, establishing a link between Butler's and Spencer's work is not an easy task. This is because direct or indirect references to Spencer and his ideas are present in all of Butler's fictional and nonfictional writings. If we consider, for example, *Evolution, Old and New* and *Luck or Cunning?*, it is possible to notice immediately that several chapters discuss or refer to Spencer's philosophy.

As with Mivart, Butler used Spencer's work in particular to support his own hypothesis concerning memory and heredity. This was because both Butler and Spencer tried to move the focus of thinking about evolution from the realm of natural science to an analysis of the influence evolution had on society and on the individual as a whole. This different approach produced a discussion of evolution in terms of psychology and then sociology for Spencer — and in relation to the mind and memory for Butler. One might broadly describe Spencer's work as an example of Social Darwinism (though strictly speaking, as will be shown, he was more a proponent of Lamarckism than Darwinism).[77] Indeed, Spencer advanced a reading of evolution that tried to mediate (like Mivart and Butler) the idea of *transformisme* with the concept of natural selection.[78]

In tracing the influence that Spencer had on Butler, we have to start with *Evolution, Old and New*. In the book, Butler's aim was to discuss and frame Spencer's ideas in relation to how evolution was explored before 1859. Indeed, although Butler in *Evolution, Old and New* included only a small number of quotations from Spencer's writing, his aim was to show how Lamarckian ideas were central to Spencer's work from the very beginning. Here, Butler referred specifically to Spencer's essay 'The Development Hypothesis', published 1852 in the periodical *The Leader*. Butler used Spencer's essay as a case study to explain how any work interested in making a contribution to the evolutionary debate could not avoid being influenced by the notion of *transformisme*. In this context, Butler cited a specific passage from Spencer's 'The Development Hypothesis': '"Those who cavalierly reject the theory of Lamarck and his followers as not adequately supported by facts," wrote Mr. Herbert Spencer, "seem quite to forget that their own theory is supported by no facts at all."'[79] The value of this citation appears immediately obvious. Butler's aim was to find in Spencer's essay a way to support his Lamarckian ideas in order to justify some of his criticism concerning the originality of Darwin's theory of natural selection. Spencer's argument offered the perfect starting point for such a task. Indeed, the first paragraph of the essay clearly highlights:

> Like the majority of men who are born to a given belief, they demand the most rigorous proof of any adverse belief, but assume that their own needs none. Here we find, scattered over the globe, vegetable and animal organisms numbering, of the one kind (according to Humboldt), some 320,000 species, and of the other, some 2,000,000 species (see Carpenter) and if to these we add the numbers of animal and vegetable species which have become extinct, we may safely estimate the number of species that have existed, and are existing, on the Earth, at not less than ten millions. Well, which is the most rational theory about these ten millions of species? Is it most likely that there have been ten millions of special creations? or is it most likely that, by continual modifications due to change of circumstances, ten millions of varieties have been produced, as varieties are being produced still?[80]

The passage quoted above shows a clear use of Lamarckian terminology ('continual modification' and 'change of circumstances'), and more broadly the influence that Lamarck had on the British evolutionary debate. Butler's choice of Spencer's essay is, therefore, instrumental to his desire to demonstrate that a concrete and working theory of evolution existed long before the publication of Darwin's work. Indeed, for Butler, Spencer's work 'leaves nothing to be desired. It is Buffon, Dr. Darwin, and Lamarck, well expressed'.[81]

In *Evolution, Old and New*, Butler also discussed the concept of the 'struggle for existence'. Although introduced by Spencer and then largely used by Darwin, the idea was actually, according to Butler, first suggested by Lamarck himself. Indeed, Butler claimed that even famous ideas like the notion of 'survival of the fittest' normally attributed to Spencer and Darwin were originally part of the evolutionary debate of the early nineteenth century. Butler explained, by quoting a passage from Lamarck, that:

> This seems to contain, and in a nutshell, as much of the essence of what Mr. Herbert Spencer and Mr. Charles Darwin have termed the survival of the fittest in the struggle for existence, as was necessary for Lamarck's purpose. To Lamarck, as to Dr. Darwin and Buffon, it was perfectly clear that the facts, that animals have to find their food under varying circumstances, and that they must defend themselves in all manner of varying ways against other creatures

which would eat them if they could, were simply some of the conditions of their existence. In saying that the surrounding circumstances — which amount to the conditions of existence — determined the direction in which any plant or animal should be slowly modified, Lamarck includes as a matter of course the fact that the 'stronger and better armed should eat the weaker,' and thus survive and bear offspring which would inherit the strength and better armour of its parents. Nothing therefore can be more at variance with the truth than to represent Lamarck and the other early evolutionists as ignoring the struggle for existence and the survival of the fittest; these are inevitably implied whenever they use the word *'circonstances'* or environment, as I will more fully show later on, and are also expressly called attention to by the greater number of them.[82]

In order to make this point even clearer, Butler borrowed a passage from Mivart's *Genesis of Species*, which suggested that:

> Mr. Spencer's theory — so Mr. Mivart tells us — and certainly that of Lamarck, whose disciple Mr. Spencer would appear to be, admits 'a certain peculiar, but limited power of response and adaptation in each animal and plant' — to the conditions of their existence. 'Such theories,' says Mr. Mivart, 'have not to contend against the difficulty proposed, and it has been urged that even very complex extremely similar structures have again and again been developed quite independently one of the other, and this because the process has taken place not by merely haphazard, indefinite variations in all directions, but by the concurrence of some other internal natural law or laws co-operating with external influences and with Natural Selection in the evolution of organic forms.'[83]

In *Unconscious Memory* Butler returned briefly to the relationship between Spencer and Lamarck. At the conclusion of the book he wrote:

> Did Mr. Herbert Spencer, for example, 'repeatedly and easily refute' Lamarck's hypothesis in his brilliant article in the *Leader*, March 20, 1852? On the contrary, that article is expressly directed against those 'who cavalierly reject the hypothesis of Lamarck and his followers.' This article was written six years before the words last quoted from Mr. Wallace; how absolutely, however, does the word 'cavalierly' apply to them![84]

To sum up, Butler's use and interpretation of Spencer's ideas, especially in *Evolution, Old and New*, was instrumental in showing the non-originality of Darwin's concept of natural selection. In doing so, Butler did not fully discuss Spencer's work, instead he merely used it in support of his own argument.

There is a second aspect to consider in looking at how Butler was influenced by the work of Spencer, specifically his philosophical interpretation of evolution. The evolutionary hypothesis of Spencer involved not only the organic kingdom, but also inorganic and super-organic interpretations of evolution. This attitude was already clear in 1864, when, in his *Principles of Biology*, Spencer claimed that 'to set forth the general truths of Biology, as illustrative of, and as interpreted by, the laws of Evolution: the special truths being introduced only so far as is needful for elucidation of the general truths'.[85] So, in order to further explore the significance of evolution, Spencer also became involved in the study of mental evolution. In the *Principle of Psychology* (1870–72), he questioned whether it was possible to think of evolution in a psychological perspective. This intention was clear from the preface of the second edition written in 1880. There, Spencer explained:

> When, in 1855, the First Edition of *The Principles of Psychology* was issued, it had to encounter a public opinion almost universally adverse. The Doctrine of Evolution everywhere implied in it, was at that time ridiculed in the world at large, and frowned upon even in the scientific world. Naturally, therefore, the work, passed over, or treated with but small respect, by reviewers, received scarcely any attention; and its contents remained unknown save to the select few. The great change of attitude towards the Doctrine of Evolution in general, which has taken place during the last ten years, has made the Doctrine of Mental Evolution seem less unacceptable; and one result has been that the leading conceptions set forth in the First Edition of this work, have of late obtained considerable currency. In France, some of them have been made known incidentally by the treatise of M. Taine, *De Intelligence* and the lucid exposition of Prof. Ribot in his *Psychologie Anglaise Contemporaine* has presented them all in a systematic form.[86]

Here, two points seem to be important. First, according to Spencer, the hypothesis of 'mental evolution' started becoming relevant only after the 1870s, especially thanks to the work of those like Ribot. Second, in France more than in England, Spencer's hypothesis became embedded in a pre-existing discussion concerning the significance of the role of the brain and memory in the process of evolution.[87]

In this respect, Butler's interpretation of Spencer's writing on psychology, evolution, and the mind became central, specifically, to his final book on science: *Luck or Cunning?*. There, Butler's aim was a simple one: to show the similarities between his theory of science of the mind ('unconscious memory') and the philosophical and psychological understanding of evolution put forward by Spencer.[88] Indeed, Butler made clear that: 'The only writer in connection with "Life and Habit" to whom I am anxious to reply is Mr. Herbert Spencer.'[89]

Luck or Cunning? focused, in particular, on Spencer's definition and use of the notion of memory. For Butler, Spencer's interpretation of memory was very close to that of Hering's (and consequently his own), although it also presented some problems. In *Principles of Psychology*, Spencer made abundant use of expressions such as 'the experience of the race' and 'accumulated experiences' — yet Butler doubted that a race could have any experience at all.[90] Spencer, in *Principles of Psychology*, defined memory as follows:

> Memory, then, pertains to all that class of psychical states which are in process of being organised. It continues so long as the organising of them continues; and disappears when the organisation of them is complete. In the advance of the correspondence, each more complex class of phenomena which the organism acquires the power of recognising is responded to at first irregularly and uncertainly; and there is then a weak remembrance of the relations. By multiplication of experiences this remembrance becomes stronger, and the response more certain. By further multiplication of experiences the internal relations are at last automatically organised in correspondence with the external ones; and so conscious memory passes into unconscious or organic memory. At the same time, a new and still more complex order of experiences is thus

rendered appreciable; the relations they present occupy the memory in place of the simpler one; they become gradually organised; and, like the previous ones, are succeeded by others more complex still.[91]

Although the act of repeating an experience proposed by Spencer was very similar to the hypothesis of acquiring habits proposed by Butler in *Life and Habit*, Spencer did not consider memory as the key factor of the hereditary process. This approach was, of course, problematic for Butler. In *Principles of Psychology*, the process of repetition became the justification of a philosophical analysis of memory which went far beyond biology and did not involve the notion of experience. In order to explain this different interpretation, Butler cited the following passage from Spencer's work:

> Just as we saw that the establishment of those compound reflex actions which we call instincts is comprehensible on the principle that inner relations are, by perpetual repetition, organised into correspondence with outer relations; so the establishment of those consolidated, those indissoluble, those instinctive mental relations constituting our ideas of Space and Time, is comprehensible on the same principle.[92]

Yet, Spencer's explanation did not satisfy Butler. In *Luck or Cunning?* he explained, instead, that 'personality and memory are the elements that constitute experience; where these are present there may, and commonly will, be experience; where they are absent the word "experience" cannot properly be used'.[93] By removing the central role of memory in the process of acquiring any experience, the whole idea of psychological evolution proposed by Spencer became, then, immediately unclear. Therefore, it is of no surprise to see how Butler came to the following conclusion:

> 'Principles of Psychology' can hardly be called clear, even now that Professor Hering and others have thrown light upon them. If, indeed, they had been clear Mr. Spencer would probably have seen what they necessitated, and found the way of meeting the difficulties of the case which occurred to Professor Hering and myself.[94]

According to the author of *Erewhon*, then, it was not possible for anyone to fully understand Spencer's argument. Because Spencer did not consider memory as central to the hereditary process, he was unable to understand and bind the force of memory as Hering had done, nor did 'he show any signs of perceiving the far-reaching consequences that ensue if the phenomena of heredity are considered as phenomena of memory'.[95] However, in expressing these concerns, Butler's goal was still to demonstrate that not only himself but also Spencer recognized the innovative importance of Hering's idea. He wrote:

> In his chapter on Memory, Mr. Spencer certainly approaches the Heringian view. He says, 'On the one hand, Instinct may be regarded as a kind of organised memory; on the other, Memory may be regarded as a kind of incipient instinct' ('Principles of Psychology,' ed. 2, vol. i. p. 445). Here the ball has fallen into his hands, but if he had got firm hold of it he could not have written, 'Instinct *may be* regarded as *a kind of*, &c.;' to us there is neither 'may be regarded as' nor 'kind of' about it; we require, 'Instinct is inherited

> memory,' with an explanation making it intelligible how memory can come to be inherited at all. I do not like, again, calling memory 'a kind of incipient instinct;' as Mr. Spencer puts them the words have a pleasant antithesis, but 'instinct is inherited memory' covers all the ground, and to say that memory is inherited instinct is surplusage.[96]

According to Butler, Spencer was, indeed, influenced by the work of Hering in the development of his psychological work, but he was not able to go as far as the German in discussing the link between memory and heredity.

In concluding this section, it is important to summarize how Butler considered Spencer's work. First, for Butler, Spencer contributed to the recognition and discussion of Lamarck's hypothesis of evolution in England especially in relation to the idea of the 'struggle for existence'. Second, Butler also recognized a similarity between the work of Spencer and Hering, focusing, in particular, on their writings concerning memory, instinct, heredity, and the science of the mind.

George Romanes and Butler's Remarks on Professional Science

The final figure to look at is George Romanes. He is particularly relevant in our examination of Butler's writing on science due to their differing opinions about mental evolution, scientific authority, and professionalism. In addition, their dissimilar personal and professional relationships with Darwin determined the professional and public reception of their theories.

In 1884, Butler published a collection of essays entitled *Selections from Previous Works and Remarks on George Romanes' Mental Evolution in Animals*. The story behind this volume is an interesting one, which can tell us something about Butler's attitude toward the work of others. In the early months of 1884, Butler was preparing a selection of his previous works for publication in a single volume, for the London editor Trübner, when George J. Romanes' *Mental Evolution in Animals* was published. After reading Romanes' book, Butler claimed that he noticed a resemblance between his idea of memory as heredity and Romanes' hypothesis of mental evolution. Butler was convinced that Romanes' theory presented certain key similarities regarding the inheritance of habits, the role of instinct, and the relationship between memory and heredity in relation to the theory he had advanced in 1878 in his first book on evolution: *Life and Habit*. Butler therefore decided to extend the scope of his *Selections from Previous Works* and include a long remark on Romanes' work and ideas. Butler's new book (from now on referred to here as the *Remarks*) presents two interesting claims. In the *Remarks*, Butler, first, tried to demonstrate how certain ideas, originally proposed by Lamarck, were used, often without crediting the naturalist, by Romanes. Second, he also tried to show a similarity between his work and Romanes' in order to make his Lamarckian reading of evolution acceptable in the eyes of professional science.

In *Life and Habit*, Butler suggested that natural selection had to be complemented by a theory of 'inheritance', which was very similar to the one proposed in 1809, by Lamarck. For Butler, heredity could only be explained as a form of organic memory. In coming to this conclusion, Butler relied heavily, as seen before, on the work of St. George Mivart as well as the idea of an 'accumulation of memory' suggested, as discussed in chapter one, in 1873 by the French psychologist and philosopher Théodule-Armand Ribot.

Butler's theory of memory as heredity was then expanded in 1880 in *Unconscious Memory*. In the volume, in order to make his theory appear more scientifically sound Butler translated into English the work of German physiologist Ewald Hering. *Unconscious Memory* was not received with much enthusiasm by the Victorian scientific community. Several reviews rejected Butler's theory as an attempt by an amateur to engage with a topic that was far too big for his own understanding. A few examples can illustrate the point: on 2 December 1880, the *St. James's Gazette* described the argument of *Unconscious Memory* as a 'rather flat' attack on Darwin supported by little biology;[97] *The Athenaeum* defined Butler's work as at best an 'ingenuous speculation'[98] and, in January 1881, a review published by *The Journal of Science* addressed the book as potentially interesting but also 'interpenetrated with polemical and personal matters' which compromised its scientific value.[99]

In 1982, the historian of science Philip J. Pauly stated that:

> [f]aithful Darwinians such as Romanes, Grant Allen, Leslie Stephen, and Fredrick Pollock all reacted strongly to Butler's attack on Darwin. They ostentatiously ignored his hypotheses regarding evolution. Instead, they claimed that he had forfeited his right to speak by reason of his lack of deference towards Darwin.[100]

Pauly explained that the main reason why loyal Darwinians ignored Butler's theory was due to the quarrel he had with Darwin.

For this reason, in 1881, Romanes publicly rejected Butler's hypothesis of unconscious evolution in his review of *Unconscious Memory*, published in *Nature*. In the review, Romanes insisted on showing how Butler's ideas did not have any scientific value because of his lack of scientific competence. Specifically on Butler's writing on science, Romanes wrote: 'To this arena, [science] however, he is in no way adapted, either by mental status or mental equipment.'[101]

Romanes' review seemed to be more focused on taking a position in the quarrel between Darwin and Butler rather than on discussing the biological implications of his Lamarckian idea of 'memory as heredity'. As explained by Pauly, Romanes did not engage with Butler's idea because the latter lacked any scientific credentials and, therefore, his work could not be valid in the eyes of the scientific community.[102] The conclusion of the review, in particular, even went beyond a personal attack by saying that Butler's work could have a value — but only among homoeopathists.[103] This was because one of the few positive reviews of *Unconscious Memory* had been published by the *Journal of Homeopathy*, which regarded Butler as an 'original thinker' who contributed to re-establishing the scientific credibility of figures like Buffon, Lamarck, and Erasmus Darwin.[104] The harsh tone of the *Nature* review led Darwin to discuss it in a letter he addressed to Romanes (28 January 1881). Darwin wrote:

> I have just read your review in Nature with the greatest interest [...] I think that you have been almost too severe. It seems to me that you have hit the right nail

on the head in attributing his conduct to the disappointment of his inordinate vanity [...] Good Lord how he will hate you. It is heroic of you to save my devoted head by calling down on your own his malignant revenge.[105]

A couple of days later, Romanes declared in his reply to Darwin: 'I cannot think that either the morality or courtesy of the scientific world is likely to be improved by the renewed exertions on their behalf which are about to be made by Mr. Samuel Butler.'[106] The two quotes above show the growing distance between Butler and Darwin both in terms of ideas and status. Nonetheless, it also expresses the distance between the amateur and the professional scientific community. Indeed, the different professional positions of Butler and Romanes made them unable to communicate with each other and also prevented Romanes from recognizing the similarities between their hypotheses.

Therefore, for Butler, showing a resemblance between his and Romanes' work became an imperative. Only in this way could Butler hope to convince the scientific community of the validity of his hypothesis of 'unconscious memory', which had been, until then, rejected as a pseudoscientific idea. So, in order to demonstrate to both the Victorian professional and lay audiences the actual scientific nature of his work, Butler made, in the *Remarks*, a very precise claim: 'The passages I have quoted show that Mr. Romanes is upholding the same opinions as Professor Hering's and my own.'[107]

Before moving on to the analysis of the *Remarks*, we must return, briefly, to Pauly's article. It highlights two interesting pieces of information about Butler and Darwinism that can help us to understand the wider context in which Butler was writing against professional science. First, Pauly suggested that Butler 'had followed the intellectual path of many Darwinians. In the late 1850s, he abandoned the clerical career planned for him by his father and converted to Darwinism while working as a gentleman sheep rancher in New Zealand'.[108] In late 1859, Butler, as discussed before, abandoned a life in the church and moved to New Zealand. Once he had arrived in New Zealand, he read Darwin's *Origins of Species* and 'became one of Mr. Darwin's many enthusiastic admirers'.[109]

Second, Pauly also observed that as a popularizer, Butler applied the theory of evolution to questions of a more general interest.[110] This approach made Butler's work difficult to be classified as scientifically accurate by his contemporaries but also later by some historians. The difficulty of framing Butler's work as 'scientific' was due to the fact that his desire to discuss evolution without conducting proper research was not well received in the 1870s when many professional scientists started to advocate their status.

Romanes' research and writings were not approached and developed in the same way. His collaboration with Darwin along with his status as a professional scientist imposed some limits to his work. However, toward the end of his career, Romanes ventured into exploring topics similar to the ones discussed by Butler, including philosophy and Lamarckism. In *Darwin's Disciple: George John Romanes, a Life in Letters* (2010), the historian Joel S. Schwartz shows how Darwin viewed Romanes as more than a young enthusiastic scientist.[111] Darwin considered Romanes as an ally. This point is central to the question that Schwartz asks in his work: 'Why was Darwin so persistent in wanting to see Romanes?'[112] Firstly, Darwin desired to meet the young scientist in order to discuss common scientific interests. However, as Schwartz writes, Darwin was also looking for a young scientist to be directed 'toward research with plants, work he believed would bring eventual success in gathering evidence in support of pangenesis, the theory of heredity he was trying to develop'.[113]

The collaboration between Darwin and Romanes began in 1874 and continued until Darwin's death. Biographers are particularly keen to define the Darwin-Romanes relationship as more than a simple professional association. After their first meeting, Darwin reputedly stated: 'How glad I am that you are so young!'[114] As Schwartz has emphasized, this positive attitude was due to Darwin's need to find a passionate collaborator as well as a future disciple. Indeed, Darwin chose Romanes to carry on his scientific research. As a result, Darwin acted as a mentor to the young scientist by introducing him to the right circles and guiding Romanes though the complexity of the evolutionary debate.

In September 1875, a year after their first acquaintance, Darwin sponsored Romanes so that he could be admitted as a fellow to The Linnean Society of London. In the same period, Darwin also placed Romanes in charge of the experiments on his hypothesis of pangenesis. The hypothesis was presented in *The Variation of Animals and Plants under Domestication* (1868) as a hypothetical mechanism for heredity. Darwin explained that pangenesis brought 'together a multitude of facts which are at present left disconnected by any efficient cause'.[115]

Darwin's hypothesis was in line with his need to deny any possible final cause in the evolutionary process. To understand Darwin's position, it helps to look at the summary of pangenesis formulated by Hugo de Vries (1848–1935) in 1889.[116] The Dutch botanist argued that Darwin's gemmules 'are much larger than the chemical molecules and smaller than the smallest known organisms [...]. They are transmitted, during cell-division, to the daughter-cells: this is the ordinary process of heredity'.[117]

Between 1874 and 1880, Romanes conducted several experiments in an attempt to verify Darwin's hypothesis of pangenesis. However, Romanes soon realized that the work on pangenesis proved to be more difficult than had been anticipated. In a letter to Darwin (14 July 1875), he explained and emphasized his intention to carry on conducting experiments to verify pangenesis, although he was not getting the expected results:

> As I am a young man yet, and hope to do a good deal of 'hammering', I shall not let Pangenesis alone until I feel quite sure that it does not admit of being any further driven home by experimental work; and even if I never get positive results, I shall always continue to believe in the theory.[118]

A few years later, in a letter to his Oxford friend Edward Bagnall Poulton (11 November 1889), Romanes admitted the failure of his research on the pangenesis hypothesis: 'Although I spent more time and trouble than I like to acknowledge (even to myself) in trying to prove Pangenesis between '73 and '80, I never obtained any positive results.'[119] Indeed, in the 1880s, August Weismann's work on germ plasm theory showed the inefficacy of the pangenesis hypothesis.[120]

Returning to the letter to Poulton, there is another point that merits attention. Romanes explained how his collaboration with Darwin also limited his freedom in conducting other types of research. In particular, Darwin tried to divert his attention away from the work of Lamarck. In the letter, Romanes explained that when he tried to propose a new investigation of Lamarck's idea of inheritance (as a possible answer to the hereditary question), Darwin dissuaded him from pursuing the matter any further. Romanes explained:

> Then it was that Darwin dissuaded me from going on to this point, on the ground that there was abundant evidence of Lamarck's principles apart from use and disuse of structures e.g. instincts and also on the ground of his theory of Pangenesis. Therefore I abandoned the matter, and still retain what may thus be now a pre-judice against exactly the same line of thought as Darwin talked me out of in 1873.[121]

In 1877, Darwin sent Romanes a selection of passages from Lamarck's *Philosophie zoologique* in order to make him aware of the difficulties presented by the work of the French natural philosopher.[122] Schwartz has suggested that Darwin wished to divert Romanes' attention away from the more philosophical aspects of Lamarck's work which had been highlighted by Spencer's interpretation.[123] For Darwin, Lamarck did not recognize the necessity of having a scientific hypothesis that was 'pure' and free from philosophical ideas. This opens up a key question regarding the role played by metaphysics in Romanes' work.

The place and significance of philosophy in Romanes' work is particularly controversial: in his main scientific publications, following Darwin's advice, he denied the role of philosophy in evolution. However, in a series of short essays he pursued his dream to bridge the gap between science and religion. This desire can be traced back in Romanes' work to 1873, with the publication of the essay 'Christian Prayer and General Laws'. On this topic, Romanes published extensively in the periodical *The Nineteenth Century*. It is important to note that *The Nineteenth Century* was edited by the founding member of the London Metaphysical Society — James Knowles (1831–1908). The periodical welcomed papers that partially mirrored the aim of the society, which attempted some intellectual rapprochement between science and religion.[124] A few examples from Romanes' own work can illustrate this point.

In 1878, in the essay 'A Candid Examination of Theism by Physicus', Romanes discussed the notion of deism as presented by John Stuart Mill (1806–1873) and William Paley. For Romanes, their work rested on an impossible metaphysical conception of nature, especially when compared to modern scientific ideas such as evolution and natural selection. Romanes wrote:

> We know by our personal experience what are our own relations to the material world, and to the laws which preside over the action of physical forces; while we can have no corresponding knowledge of the relations subsisting between the Deity and these same objects of our own experience. Hence, to suppose that the Deity constructed the eye by any such process of thought as we know that

> men construct watches, is to make an assumption not only incapable of proof, but destitute of any assignable degree of likelihood.[125]

However, after rejecting the ideas of Mill and Paley, Romanes arrived at the unexpected conclusion 'that the hypothesis of metaphysical teleology, although in a physical sense gratuitous, may be in a psychological sense legitimate'.[126] This point is extremely significant, especially when read in relation to Romanes' letter to Poulton and Lamarckism.

There is no need to present Romanes' work as a form of Lamarckism, but it is undeniable that in contrast to the teaching of Darwin, Romanes, towards the end of his career, tried to give teleology a second chance in evolution. In the essay 'The Fallacy of Materialism' (1882) Romanes took a position in favour of a partial return to the idea of design. He was particularly intrigued by the possibility of solving the gap between science and religion via spiritualism. What is surprising is that even in *Darwin and After-Darwin*, Romanes wrote:

> [The change] has done nothing in the way of negativing that belief in a Supreme Being in which it was the object of these authors [Paley, Bell, Chalmers] to substantiate. If it has demonstrated the futility of their proof, it has furnished nothing in the way of disproof. It has shown, indeed, that their line of argument was misjudged when they thus sought to separate organic nature from inorganic as a theatre for the special or peculiar display of supernatural design; but further than this it has not shown anything.[127]

A few pages later, Romanes even declares that

> [h]e [Darwin] has proved, more rigidly than was ever proved before, that suffering is a condition to improvement — struggle for life being the raison d'être of higher life, and this not only in the physical sphere, but also in the mental and moral.[128]

In a paper entitled 'Evidences of Design in Nature' presented at the Aristotelian Society in 1890, Romanes explained that '[t]here is nothing in the constitution of nature inimical to the hypothesis of design'.[129] However, Romanes also insisted that 'science has reduced the great and old-standing question of Design in Nature to this comparatively narrow issue'.[130] What is important to note here is that for Romanes evolution could not be completely separated from philosophy. Therefore, it is easy to see how he was unable to completely reject Lamarckism. Romanes' reference to Lamarckism is not surprising, particularly when it is examined in relation to Romanes' essays in *The Nineteenth Century*. The philosophical response to evolution embraced by Romanes ventured into a territory not that different from that explored by Butler. The main difference between the two thinkers was only determined by their different professional positions and their consequent approach to science.

In a letter written to his friend and long-time correspondent Ann Savage on 20 January 1884, Butler explained the origin of the *Remarks* as follows:

> He [Romanes] has calmly cribbed my Life and Habit theory — of course without acknowledgement, so I have put in a chapter after the extracts from *Life and Habit*, *Evolution Old and New*, and *Unconscious Memory*, quoting the passages

in his book in which he says what I have been saying, and chaffing him for the way in which he abused this very theory not three years since.[131]

Butler claimed that Romanes' hypothesis of mental evolution was very similar to Hering's notion of memory and heredity, and that Romanes did not give the work of the German thinker a proper acknowledgement. Butler was not the only one to have noted this similarity. In a review of *Mental Evolution in Animals*, published in *The Athenaeum* in 1884, the psychologist and philosopher James Ward (1843–1925) accused Romanes of not having sufficient scientific ground to attack Butler while adopting similar ideas (without acknowledging it).[132] Ward, while recognizing Butler's status as an amateur, made clear that important developments in Darwin's evolutionary theory had been advanced by non-professionals like Grant Allen and Herbert Spencer. Therefore, Romanes 'practical' acceptance of Butler's theory showed that even an amateur could do a service to science. However, this service, Ward wrote, must also be acknowledged. The philosopher pointed out that '[t]he phrase "Hereditary Memory" is due to Mr Samuel Butler author of "Erewhon" and "Life and Habit"' and suggested considering Butler's work as an important contribution to the evolutionary debates and not just a nonsense theory.[133]

In the *Remarks*, Butler tried to illustrate Ward's argument. Butler began by providing an overview of how 'memory' was deployed by Herbert Spencer in the *Principles of Psychology* (instinct as organized memory); by the philosopher George Henry Lewes (1817–1878) in the 1873 *Problems of Life and Mind* (lapsing of intelligence), and finally, by Romanes — predominately as a form of instinct (before *Mental Evolution in Animals*).[134] In contrast to their definitions, memory, in Butler's view, had not only to be considered as the key factor for understanding evolution but also as both a physical and metaphysical justification of human life.

What is important here is that Butler recognized that, unlike Hering, neither Spencer nor Lewes, nor Romanes were able to understand the importance of memory in evolution. This incapacity to fully embrace the importance of memory was because their definitions of memory, Butler explained, did not take into account the recent development of the debate in continental Europe.[135] Butler wrote:

> none of these writers (and indeed no writer that I know of except Professor Hering of Prague, for a translation of whose address on this subject I must refer the reader to my book *Unconscious Memory*) has shown a comprehension of the fact that these expressions are unexplained so long as 'heredity,' whereby they explain them, is unexplained; and none of them sees the importance of emphasizing Memory, and making it as it were the keystone of the system.[136]

As discussed in chapter one, in 1870 Hering delivered his famous lecture at the University of Prague. The lecture rapidly became one of the most frequently quoted texts by both psychologists and physiologists. It gave rise to a series of translations and was largely used among European physiologists. In terms of content, Hering's paper identified memory as a fundamental reproductive capability of living matter. He recognized that '[i]t is to memory [...] that we owe almost all that we have or are; our ideas and conceptions are its work; our every thought and movement are derived from this source'.[137] The main hypothesis put forward in Hering's lecture was the necessity to link materialistic science (physiology) with the philosophy of the mind (psychology). Hering's study involved a deep scientific examination of the correlation between memory, heredity, and Lamarckism while also discussing philosophical problems.

In the *Remarks*, Butler, citing *Unconscious Memory*, suggested that memory is the condition that connects all (material and immaterial) aspects of human existence into a single whole. To explain this, Butler pointed out that 'bodies would be scattered into the dust of their component atoms if they were not held together by the cohesion of matter';[138] in the same way, Butler clarified, that 'our consciousness would be broken up into as many moments as we had lived seconds, but for the binding and unifying force of Memory'.[139] Butler came to the conclusion that Romanes in

> his recent work, *Mental Evolution in Animals*, shows that he is well aware of the direction which modern opinion is taking, and in several places he so writes as to warrant me in claiming his authority in support of the views which I have been insisting on for several years past.[140]

By saying this, Butler claimed that Romanes' hypothesis of mental evolution was more similar to his hypothesis of memory as heredity than to Darwin's position.

This opens up the central point of the *Remarks*. In *Mental Evolution in Animals* Romanes defined memory as distinguishing between mechanical actions, where the individual has no control, and conscious actions that are based on the individual's choice and desires. Romanes wrote: 'The distinctive element of mind is consciousness, the test of consciousness is the presence of choice, and the evidence of choice is the antecedent uncertainty of adjustive action between two or more alternatives.'[141] The quotation is beneficial to Butler's argument, which was aimed at demonstrating how Romanes adopted concepts and terms very similar to the language used by Hering, and by himself in *Life and Habit*.

It is important to take a look at the terminology and examples employed by Romanes with particular attention to how he refers to the notion of inheritance. In order to demonstrate the validity of his argument, Butler cited several passages of Romanes' *Mental Evolution in Animals*. Butler explained that Romanes defined heredity in relation to the perceptive faculty of the individual prior to its own experience.[142] Romanes had written: 'that automatic actions and conscious habits may be inherited'[143] and even claimed that instinct 'may be lost by disuse, and conversely that they may be acquired as instincts by the hereditary transmission of ancestral experience'.[144]

Additionally, in order to explain this 'hereditary transmission' as a form of inherited memory, Romanes provided the example of a migratory bird, which 'possess[es], by inheritance alone, a very pre knowledge of the particular direction to be pursued'.[145] Romanes claimed that beyond question any young bird knows in which particular time of the year to leave its parents, and without any guide how to arrive at its migratory spot.[146] This phenomenon, Romanes concluded, can only be demonstrated 'by taking it to be due to inherited memory'.[147]

The example of the bird is useful in showing how repetitions and habit were becoming key aspects of Romanes' idea of memory.[148] Therefore, it is not surprising that Romanes arrived at the conclusion that: 'Now upon our own theory it can only be met by taking it to be due to inherited memory.'[149] This idea was then explained further in this passage:

> That 'practice makes perfect' is a matter, as I have previously said, of daily observation. Whether we regard a juggler, a pianist, or a billiard-player, a child learning his lesson or an actor his part by frequently repeating it, or a thousand other illustrations of the same process, we see at once that there is truth in the cynical definition of a man as a 'bundle of habits.' And the same of course is true of animals.[150]

In *Life and Habit*, Butler used a similar array of examples to explain the relationship between memory, habit, and heredity (I quote the full passage to illustrate the point):

> Taking then, the art of playing the piano as an example of the kind of action we are in search of, we observe that a practised player will perform very difficult pieces apparently without effort, often, indeed, while thinking and talking of something quite other than his music; yet he will play accurately and, possibly, with much expression. If he has been playing a fugue, say in four parts, he will have kept each part well distinct, in such a manner as to prove that his mind was not prevented, by its other occupations, from consciously or unconsciously following four distinct trains of musical thought at the same time, nor from making his fingers act in exactly the required manner as regards each note of each part.[151]

Therefore, Butler, in his *Remarks*, was able to come to the following conclusion:

> There can be no doubt, however, that Mr. Romanes does in reality, like Professor Hering and myself, regard development, whether of mind or body, as due to memory, for it is nonsense indeed to talk about 'hereditary experience' or 'hereditary memory' if anything else is intended.[152]

Consequently, Butler's *Remarks*, instead of critiquing Romanes' ideas, became the opportunity to show how the hypothesis of mental evolution had more in common with Hering's idea of memory as heredity than with Darwin's concept of natural selection.

In concluding this section there is a final point that remains to be explored. Towards the end of the *Remarks*, Butler said that Romanes — in *Mental Evolution in Animals* — had claimed that Charles Kingsley was the first to advance the hypothesis of memory as heredity in an article published in *Nature* on 18 January 1867.[153] Butler, of course, did not consider this claim correct. As observed by Schacter, in *Evolution Old and New* Butler pointed out that a basic interpretation of memory as heredity had previously been advanced by Patrick Matthew in the 1831 book *On Naval Timber and Arboriculture*.[154] In the same book, Butler also named Matthew alongside Lamarck and Erasmus Darwin as the 'founders of the theory' that up until that point only Hering and Butler himself had insisted on.[155]

Additionally, in a letter sent to *The Athenaeum* (26 January 1884) Butler claimed that:

Nature did not begin to appear till nearly three years after the date given by Mr. Romanes, and that there was nothing from Canon Kingsley on the subject of instinct and inherited memory in any number of Nature up to the date of Canon Kingsley's death.[156]

On this matter, more information can be found in Butler's correspondence with his sister May. On 2 February 1884, Butler wrote:

> He [Romanes] said Canon Kingsley was the first to advance the theory connecting heredity and memory, in Nature, January 18th, 1867. I wrote and pointed out that Nature did not exist till November, 1869, and that Canon Kingsley never said anything in Nature at all, in any way bearing on memory and heredity and asked for the right reference. Romanes did not answer — so I have given it him hot. We don't think Canon Kingsley ever said anything at all about it.[157]

This is an interesting historical misunderstanding: Kingsley, of course, did not publish in *Nature*. However, in 1867, he wrote a short article in *Fraser's Magazine for Town and Country* entitled 'The Charm of Birds'. In the article, Kingsley explained that the wood wren, a bird of the British Channel Islands

> felt, nevertheless, that 'that was water he must cross,' he knew not why: but something told him that his mother had done it before him, and he was flesh of her flesh, life of her life, and had inherited her 'instinct' (as we call hereditary memory, in order to avoid the trouble of finding out what it is, and how it comes).[158]

Although Kingsley used the word hereditary memory, he did not develop the point in the article or in any other of his books or essays.

The first reference to Hering and the notion of memory and heredity was discussed by Ray Lankester in an article published in *Nature* in 1876: 'Perigenesis v. Pangenesis: Haeckel's New Theory of Heredity'. Here, the biologist briefly mentioned the name of Hering but without providing a comprehensive summary of his idea. The first full account of Hering's work was, then, provided by Butler in *Unconscious Memory*. Therefore, it does not come as a surprise that in the *Remarks* Butler was keen to point out:

> For the present it is enough to say that if he [Romanes] does not mean what Professor Hering and, longo intervallo, myself do, he should not talk about habit or experience as between successive generations, and that if he does mean what we do — which I suppose he does — he should have said so much more clearly and consistently than he has.[159]

Butler and Romanes proposed, in their respective works, a similar interpretation of memory in evolution. However, because of their different professional positions and relationships with Darwin, only Romanes had the chance to have his work recognized by the Victorian scientific community. In contrast, Butler, despite all of his efforts in popularizing European ideas in Britain, became known to contemporaries as the thinker who in relation to science was 'in no way adapted, either by mental status or mental equipment'.[160]

Notes to Chapter 4

1. Butler, *Evolution, Old and New*, pp. 60–61.
2. The popularization of science in the Victorian period has been the object of many recent historical studies. These included, but are not limited to, *Victorian Science in Context* (ed. by Bernard Lightman, 1997), Aileen Fyfe's *Science and Salvation* (2004), Lightman's *Victorian Popularizers of Science* (2007), and *Science in the Marketplace* (ed. by Aileen Fyfe and Bernard Lightman, 2007). All of these studies contributed, in different ways, to providing a new scholarly significance to terms like 'popular' and 'popularization' and the role they played in the development of the Victorian scientific discourse. To have an overview of some of the topics and for a general introduction see Bernard Lightman, '"The Voices of Nature": Popularising Victorian Science', in *Victorian Science in Context*, ed. by Bernard Lightman (Chicago: Chicago University Press, 1997), pp. 187–211; Aileen Fyfe, *Science and Salvation: Evangelical Popular Science Publishing in Victorian Britain* (Chicago: University of Chicago Press, 2004).
3. Fyfe, *Science and Salvation*, p. 56.
4. Fyfe explains that 'the existence of a "mass audience" was just beginning to be recognised in the 1840s, and it was frequently perceived as a crowd of different sorts of people rather than as the homogeneous mass that we tend to think of today'. See Fyfe, *Science and Salvation*, p.6.
5. Both *Victorian Science in Context* and *Science and Salvation* suggest that the popularization of science in the Victorian period was a phenomenon existing somewhere in between the popular audience and the professionalism of science. Therefore, in further exploring the meaning and significance of Butler's writing on science we need to look at his work in that particular light: in between the activities and existence of amateurs and professionals.
6. There is a large scholarship on Victorian periodicals. See Geoffrey Cantor and others, *Science in the Nineteenth-Century Periodical: Reading the Magazine of Nature* (Cambridge: Cambridge University Press, 2008); *Science Serialized*, ed. by Cantor and Shuttleworth; the project *Science in the Nineteenth-Century Periodical* <http://www.sciper.org/> [accessed 9 January 2020].
7. *Science in the Marketplace: Nineteenth-Century Sites and Experiences*, ed. by Aileen Fyfe and Bernard Lightman (Chicago: University of Chicago Press, 2007).
8. Jonathan Topham, 'Publishing "Popular Science" in Early Nineteenth-Century Britain', in *Science in the Marketplace*, ed. by Fyfe and Lightman, pp. 135–68 (p. 159).
9. Topham, 'Publishing "Popular Science"', p. 136.
10. Topham, 'Publishing "Popular Science"', p. 136.
11. See James Secord, 'How Scientific Conversation Became Shop Talk', *Science in the Marketplace*, ed. by Fyfe and Lightman, pp. 23–59 (pp. 34–35).
12. Secord, 'How Scientific Conversation became Shop Talk', p. 35.
13. Secord, *Victorian Sensation*, pp. 32–34, 131.
14. Aileen Fyfe and Bernard Lightman, 'Science in the Marketplace: An Introduction', in *Science in the Marketplace*, ed. by Fyfe and Lightman, pp. 1–19 (pp. 10–12).
15. White, *Thomas Huxley*, pp. 32–58.
16. See Ruth Barton, '"Huxley, Lubbock, and Half a Dozen Others": Professionals and Gentlemen in the Formation of the X Club, 1851–1864', *Isis*, 89.3 (1998), 410–44, as well as her '"An Influential Set of Chaps": The X-Club and Royal Society Politics 1864–85', *The British Journal for the History of Science*, 23.1 (1990), 53–81; James Desmond, 'Redefining the X Axis: "Professionals", "Amateurs" and the Making of Mid-Victorian Biology: A Progress Report', *Journal of the History of Biology*, 34.1 (2001), 3–50; and White, *Thomas Huxley*, p. 98.
17. Lightman, *Victorian Popularizers of Science*, pp. 39–95.
18. Lightman, '"A Conspiracy of One"', pp. 113–42.
19. It is my contention that Samuel Butler represents a case which is not fully compatible with the model of the marketplace as defined by previous scholarship. First, Butler's 'marketplace' was different from that of his peers as — to maintain the marketplace analogy — he bought and even sold beyond the local stalls (e.g. the British and European markets); Butler's writing on evolution, as a critique of Darwinian authority and late nineteenth-century professionalism, was based on the work of French and German natural historians, physiologists, psychologists, and philosophers. Second, Butler's science of the mind was not just a commentary on evolution

but rather a genuinely speculative theory based on empirical research conducted in continental Europe.
20. Butler, *Life and Habit*, p. 2.
21. White, *Thomas Huxley*, pp. 51–58.
22. White, *Thomas Huxley*, p. 5.
23. Lightman, *Victorian Popularizers of Science*, p. 13.
24. Lightman, *Victorian Popularizers of Science*, pp. 35–37.
25. Lightman, '"A Conspiracy of One"', pp. 118–20.
26. Lightman, '"A Conspiracy of One"', pp. 131–33, 138.
27. Schacter, *Forgotten Ideas, Neglected Pioneers*, pp. 113–19.
28. Butler, *Life and Habit*, p. 35.
29. For instance, in *Life and Habit* Butler cited the work of philosophers like Marcus Aurelius or even religious figures such as St Paul and discussed several passages from the Bible whilst explaining his ideas regarding memory, habits, and instinct.
30. Butler, *Life and Habit*, pp. 1–2.
31. Butler, *Luck or Cunning?*, p. 26.
32. Butler, *Luck or Cunning?*, p. 27.
33. *Life and Habit*, for example, includes a large number of quotations. In reading the book, a Victorian could find references to Charles Darwin, Wallace, Mivart, and Huxley, but, and at the same time, also to the work of French authors like Claude Bernard, Théodule Ribot, and Pierre Huber (1777–1840). Besides contemporary theories published in England and abroad, Butler also largely cited pre-Darwinian naturalists including Lamarck, Buffon, Erasmus Darwin, and Isidore Geoffroy Saint-Hilaire.
34. Butler, *Life and Habit*, p. 2.
35. It is important to highlight how the work of translation contributed to the flourishing of both science and culture during the nineteenth century despite being also very controversial in some instances. A famous and extreme example, which illustrates the importance played by translation in the circulation of scientific ideas, was the 1862 translation into French of the *Origin of Species*. It is sufficient to look at the title to understand how the translator modified the text to his own will. The original title *On the Origin of Species by Means of Natural Selection, or the Preservation of Favoured Races in the Struggle for Life* became in the French translation *De l'Origine des espèces, ou Des Lois du progrès chez les êtres organisés*, meaning that vital concepts like 'natural selection' and 'struggle for life' disappeared in the French edition. Additionally, the preface of the volume, written by the translator Clémence Royer, presented Darwin's work as a form of progressive evolution which had more in common with the ideas of Lamarck than with those of Darwin. This different title was chosen because in France the idea of natural selection, as the main explanation of the evolutionary process, was largely refuted by national scientific institutions. For instance, *L'Academie des Sciences*, the French version of the Royal Society, rejected Darwin and his idea of natural selection for the first time in 1870 and definitively in 1878. Darwin's theory of evolution began to be accepted in France only in the early twentieth century. See Joy Harvey, *Almost a Man of Genius: Clémence Royer, Feminism and Nineteenth-Century Science* (New Brunswick: Rutgers University Press, 1997), p. 79; Robert E. Stebbins, 'France', in *The Comparative Reception of Darwinism*, ed. by Thomas E. Glick (Chicago: Chicago University Press, 1988), pp. 117–67; Barsanti, *Una lunga pazienza cieca*, pp. 306–08.
36. *Translators through History*, ed. by Jean Delisle and Judith Woodsworth (Philadelphia: John Benjamins Publishing Company, 1995), p. 191.
37. Butler, *Unconscious Memory*, p. 42.
38. Butler, *Unconscious Memory*, p. 63.
39. An account of the scientific importance of Mivart's work can be found in: Francesca Bigoni and Giulio Barsanti, 'Evolutionary Trees and the Rise of Modern Primatology: The Forgotten Contribution of St. George Mivart', *Journal of Anthropological Sciences*, 89 (2011), 93–107. In addition, Mark S. Burrows examines Mivart's idea concerning science and religion in his article 'A Historical Reconsideration of Newman and Liberalism: Newman and Mivart on Science and the Church', *Scottish Journal of Theology*, 40.3 (1987), 399–419.

40. George Mivart, 'Evolution and its Consequences: A Reply to Professor Huxley', *The Contemporary Review*, 19, (December 1871/May 1872), pp. 168–97.
41. Mivart, 'Evolution and its Consequences', pp. 168–70.
42. Adrian Desmond, *Archetypes and Ancestors: Palaeontology in Victorian London* (Chicago: University of Chicago Press, 1984), pp. 137–42.
43. A discussion of the popular reception of Mivart's work on evolution can be found here: Emma E. Swain, 'St. George Mivart as Popularizer of Zoology in Britain and America, 1869–1881', *Endeavour*, 41.4 (2017), 176–91.
44. George Mivart, *On the Genesis of Species* (London: Macmillan & Co., 1871), p. 4.
45. Mivart, *On the Genesis of Species*, p. 5.
46. Mivart, *On the Genesis of Species*, p. 5.
47. Mivart, *On the Genesis of Species*, pp. 5–6.
48. Thomas H. Huxley, 'Mr Darwin's Critics', *Contemporary Review*, 18 (1871), 442–76 (p. 444).
49. Thomas H. Huxley, *Darwiniana* (London: Appleton, 1893), p. 69.
50. Huxley, *Darwiniana*, pp. 70–71.
51. Charles Darwin, *On the Origin of Species by Means of Natural Selection, or the Preservation of Favoured Races in the Struggle for Life*, 6th edn, with additions and corrections (London: Murray, 1872), p. 133.
52. Darwin, *On the Origin of Species*, p. 176.
53. Darwin, *On the Origin of Species*, p. 177.
54. Darwin, *On the Origin of Species*, p. 204.
55. In the case of Mivart, the quarrel started after the publication of *The Genesis of Species*.
56. Mivart, *On the Genesis of Species*, p. 6.
57. Mivart, *On the Genesis of Species*, pp. 4–5.
58. The last chapter of Mivart's *Genesis of Species* is particularly relevant to Butler. There, Mivart discussed topics including: the work and objections of Herbert Spencer to theism; the significance and meaning of the term 'creation'; the place and role of Darwin's 'natural selection' in the evolutionary debate; the correlation and relationship of Christianity and evolution; an explanation of how theological authority is not opposed to evolution; the work of several fathers of the church including St Augustin and St Thomas; and, finally, an examination of the parallels between Christianity and natural theology looking at the work of Louis Agassiz (1807–1873), Huxley, Richard Owen (1804–1892), Darwin, and Wallace.
59. In *Life and Habit*, Butler writes: 'Mr. Mivart urges with much force the difficulty of starting any modification on which "natural selection" is to work, and of getting a creature to vary in any definite direction.' See Butler, *Life and Habit*, p. 277.
60. Butler, *Life and Habit*, pp. 179–80.
61. Butler, *Life and Habit*, p. 280.
62. Butler, *Life and Habit*, pp. 280–81.
63. Butler, *Life and Habit*, p. 284.
64. Butler, *Life and Habit*, p. 290.
65. Butler, *Unconscious Memory*, pp. 22–23.
66. Butler, *Unconscious Memory*, pp. 54–55.
67. Butler, *Unconscious Memory*, p. 55.
68. Butler, *Luck or Cunning?*, p. 16.
69. Butler, *Luck or Cunning?*, p. 19.
70. Butler, *Luck or Cunning?*, p. 49.
71. See Barton, '"Huxley, Lubbock, and Half a Dozen Others"', pp. 411, 417, 421–23.
72. See Herbert Spencer, 'Progress: Its Law and Cause', in Herbert Spencer, *Essays, Scientific, Political, and Speculative* (New York: Appleton, 1892), pp. 8–63.
73. See Samuel Taylor Coleridge, *Hints towards the Formation of a More Comprehensive Theory of Life* (London: John Churchill, 1818). An overview of Karl Ernst Ritter von Baer's theory of embryological development is available in English translation by Huxley. See Thomas Henry Huxley, *Scientific Memoirs: Selected from the Transactions of Foreign Academies of Science, and from Foreign Journals* (London: Taylor and Francis, 1853), pp. 176–238.

74. Spencer, 'Progress: Its Law and Cause'.
75. Mike Hawkins, *Social Darwinism in European and American Thought, 1860–1945: Nature as Model and Nature as Threat* (Cambridge: Cambridge University Press, 2008), pp. 83–85.
76. Herbert Spencer, *An Autobiography*, 2 vols (New York: Appleton, 1904), I, 502.
77. See the now classic Richard Hofstadter, *Social Darwinism in American Thought* (Boston: Beacon Press, 1944).
78. See Giorgio Lanaro, *L'evoluzione, il progresso e la società industriale: Un profilo di Herbert Spencer* (Roma: La Nuova Italia Editrice, 1997), pp. 93–100.
79. Butler, *Evolution Old and New*, p. 330.
80. Herbert Spencer, 'The Haythorne Papers, No. 2: The Development Hypothesis', *The Leader*, 3.104, 20 March 1852, pp. 280–81 (p. 280).
81. Butler, *Evolution, Old and New*, p. 332.
82. Butler, *Evolution, Old and New*, pp. 281–82.
83. Butler, *Evolution, Old and New*, p. 343.
84. Butler, *Unconscious Memory*, p. 183.
85. Herbert Spencer, *Principles of Biology* (London: Williams and Norgate, 1864), p. iii.
86. Herbert Spencer, *Principles of Psychology* (London: Longman, Brown, Green and Longmans, 1855), pp. iii–iv.
87. Laurent Mucchielli, 'Aux Origines de la psychologie universitaire en France (1870–1900): Enjeux intellectuels, contexte politique, réseaux et stratégies d'alliance autour de la *Revue philosophique* de Théodule Ribot', *Annals of Science*, 55 (1998), 263–89 (p. 266).
88. Butler, *Luck or Cunning?*, p. 24.
89. Butler, *Luck or Cunning?*, p. 24.
90. Spencer, *Principles of Psychology*, pp. 126, 518, 538.
91. Spencer, *Principles of Psychology*, p. 563; as quoted in Butler, *Luck or Cunning?*, p. 29.
92. Spencer, *Principles of Psychology*, p. 579; as quoted in Butler, *Luck or Cunning?*, p. 29.
93. Butler, *Luck or Cunning?*, p. 30.
94. Butler, *Luck or Cunning?*, pp. 39–40.
95. Butler, *Luck or Cunning?*, p. 49.
96. Butler, *Luck or Cunning?*, pp. 46–47.
97. Review of Samuel Butler, *Unconscious Memory: A Comparison Between the Theory of Dr. Ewald Hering, Professor of Physiology in the University of Prague, and the 'Philosophy of the Unconscious' of Dr. Edward von Hartmann, with Translations from both these Authors, and Preliminary Chapters Bearing upon 'Life and Habit', 'Evolution, Old and New', and Mr. Charles Darwin's Edition of Dr. Krause's 'Erasmus Darwin'*, in *St. James's Gazette* (2 December 1880), 13.
98. Review of Samuel Butler, *Unconscious Memory [...]*, in *The Athenaeum*, 2773 (18 December 1880), 810.
99. Review of Butler, *Unconscious Memory [...]*, in *The Journal of Science*, 3 (1881), 4.
100. Pauly, 'Samuel Butler and his Darwinian Critics', p. 163.
101. George J. Romanes, 'Mr. Butler's *Unconscious Memory*', in *Nature*, 23.587 (1881), 286–87.
102. Pauly, 'Samuel Butler and his Darwinian Critics', pp. 173–74.
103. Romanes, 'Mr. Butler's *Unconscious Memory*', pp. 286–87.
104. Review of Samuel Butler, *Unconscious Memory [...]*, in *The British Journal of Homeopathy*, 155 (1880), 67–73 (p. 71).
105. Quoted after Joel S. Schwartz, 'George John Romanes's Defence of Darwinism: The Correspondence of Charles Darwin and his Chief Disciple', *Journal of the History of Biology*, 28.2 (1995), 281–316 (p. 312).
106. Quoted after Schwartz, 'George John Romanes's Defence of Darwinism', p. 313
107. Butler, *Selections from Previous Works*, p. 236.
108. Pauly, 'Samuel Butler and his Darwinian Critics', p. 165.
109. Butler, *Unconscious Memory*, p. 11.
110. Pauly, 'Samuel Butler and his Darwinian Critics', p. 165.
111. Joel S. Schwartz, *Darwin's Disciple: George John Romanes, a Life in Letters* (Philadelphia: American Philosophical Society, 2010).

112. Schwartz, 'George John Romanes's Defence of Darwinism', p. 288.
113. Schwartz, 'George John Romanes's Defence of Darwinism', p. 288
114. Ethel Romanes, *The Life and Letters of George John Romanes* (London: Longmans, Green & Co., 1898), p. 14.
115. Darwin, *The Variation of Animals and Plants*, p. 357.
116. K. Holterhoff, 'The History and Reception of Charles Darwin's Hypothesis of Pangenesis', *Journal of the History of Biology*, 47 (2014), 661–95.
117. Hugo de Vries, *Intracellular Pangenesis* (Chicago: Chicago University Press, 1910), p. 63.
118. Romanes, *The Life and Letters of George John Romanes*, p. 42.
119. Romanes, *The Life and Letters of George John Romanes*, p. 23.
120. Zeller, Peter, *Romanes: Un discepolo di Darwin alla ricerca delle origini del pensiero* (Rome: Armando Editore, 2007), p. 48.
121. Romanes, *The Life and Letters of George John Romanes*, p. 223.
122. Schwartz, 'George John Romanes's Defence of Darwinism', p. 299.
123. Schwartz, 'George John Romanes's Defence of Darwinism', p. 299.
124. C. Hajdenko-Marshall, 'Believing after Darwin: The Debates of the Metaphysical Society (1869–1880)', *Cahiers victoriens et édouardien online*, 76 (2012), 69–83.
125. George J. Romanes, *A Candid Examination of Theism by Physicus* (Boston: Houghton, Osgood & Co., 1878), p. 37.
126. Romanes, *A Candid Examination of Theism*, p. 110.
127. George J. Romanes, *Darwin and After-Darwin: An Exposition of the Darwinian Theory and a Discussion of Post-Darwinian Questions*, 3 vols (Chicago: The Open Court Publishing Company, 1892), I, 413.
128. Romanes, *Darwin and After-Darwin*, I, 415.
129. Romanes, *The Life and Letters of George John Romanes*, p. 250.
130. Romanes, *The Life and Letters of George John Romanes*, p. 263.
131. Butler, *Letters between Samuel Butler and Miss E. M. A. Savage*, p. 319.
132. James Ward, 'Review of Mental Evolution in Animals by G. J. Romanes', *The Athenaeum*, 2773 (1884), 282–83.
133. Ward, 'Review of Mental Evolution in Animals', p. 282.
134. Butler, *Selections from Previous Works*, pp. 228–29.
135. Butler, *Selections from Previous Works*, p. 228.
136. Butler, *Selections from Previous Works*, pp. 228–29.
137. Butler, *Unconscious Memory*, p. 116.
138. Butler, *Unconscious Memory*, p. 116.
139. Butler, *Unconscious Memory*, p. 116.
140. Butler, *Selections from Previous Works*, p. 231.
141. George J. Romanes, *Mental Evolution in Animals: With a Posthumous Essay on Instinct by Charles Darwin* (London: Paul, Trench & Co. 1883), p. 18.
142. Romanes, *Mental Evolution in Animals*, p. 131.
143. Romanes, *Mental Evolution in Animals*, p. 296.
144. Romanes, *Mental Evolution in Animals*, p. 192.
145. Butler, *Selections from Previous Works*, p. 234; Romanes, *Mental Evolution in Animals*, p. 289.
146. Romanes, *Mental Evolution in Animals*, p. 289.
147. Butler, *Selections from Previous Works*, p. 234; Romanes, *Mental Evolution in Animals*, p. 289.
148. Romanes, *Mental Evolution in Animals*, pp. 177–78.
149. Romanes, *Mental Evolution in Animals*, p. 192.
150. Romanes, *Mental Evolution in Animals*, p. 193.
151. Butler, *Selections from Previous Works*, p. 68.
152. Romanes, *Mental Evolution in Animals*, p. 238.
153. As highlighted by Butler, Romanes explained that the theory of inherited memory was 'first advanced by Canon Kingsley (Nature, Jan. 18, 1867), and had since been independently suggested by several writers'. See Butler, *Selections from Previous Works*, p. 296.
154. Schacter, *Forgotten Ideas, Neglected Pioneers*, p. 115.
155. Butler, *Evolution, Old and New*, p. 382.

156. Butler, *Selections from Previous Works*, p. iii.
157. Daniel Howard, *The Correspondence of Samuel Butler with his Sister May* (Berkeley: University of California Press, 1962), p. 115.
158. Charles Kingsley, 'The Charm of Birds', *Fraser's Magazine for Town and Country*, 75 (1867), 802–10 (p. 808).
159. Butler, *Selections from Previous Works*, p. 254.
160. Romanes, 'Mr. Butler's *Unconscious Memory*', p. 285.

CONCLUSION

Another Story to Tell

> Argument is generally a waste of time and trouble. It is better to present one's opinion and leave it to stick or no [sic] as it may happen. If sound, it will probably in the end stick, and the sticking is the main thing.
>
> SAMUEL BUTLER, *The Notebooks*[1]

As stated by Stanley Bates Harkness' *The Career of Samuel Butler*, Butler had two lives: the first in England, and then between 1890 (the year when Butler published his final essay on evolution: 'The Deadlock of Darwinism') and the 1920s, in France and Italy.[2] In these two countries, Butler's work became known, discussed, and critically acclaimed by both writers and professional scientists. Although translations of Butler's main books were also made in Spanish, Dutch, French, German, and Russian, it is in Italy, and partially in France, where we can see his work really flourishing with fame.[3] Indeed, when we look at the reception of Butler's ideas and persona on the continent, there is a very different story to tell.

In this book, we have explored the British side of the story, where Butler was the outsider whose work was ignored and dismissed by the English professional community as well as by the general public. However, it is now time to shed some light on the reasons why his ideas and approach were received so differently in mainland Europe. This is the story of an unexpected reception, especially in France and Italy, where Butler's work was recognized for its literary, artistic, and scientific achievements. This contrast between his homeland and the continent is worth further examination and can help us to reframe his work and the rise and fall of his fame in a wider pan-European perspective.

Italy had always been a place of inspiration for Butler. In most of his books it is possible to find references to Italian culture and traditions from the north to the south of the peninsula. As is widely known, travelling in Italy was an important part of the life of many Victorian writers. Italy was considered a privileged place for studying art, writing, and experiencing a different lifestyle.[4] In 1920, Henry Festing Jones' *Memoir* dedicated an extensive part of the volumes to exploring the relationship between Samuel Butler and Italy. Since his childhood, Butler had the chance to travel in Italy with his family several times, and this greatly influenced his future. In addition, Jones also stressed how from the beginning of Butler's painting career, in the 1860s, the engagement with Italian culture and art became, year after year, more and more central to his work.[5]

There is no need to go into much detail here but if we take, as an example, *The Way of all Flesh*, we can see how the main character Ernest Pontifex visited Italy during his childhood and returned there several times later on in his life. Exactly like Butler, the young Ernest spent time exploring the peninsula, embracing its culture, and learning the local language.[6] It is also no surprise to find so many allusions to the country in *Erewhon*. Although the novel is set in the New Zealand countryside, the physical appearance of its population recalls Mediterranean traits. Indeed as Butler writes in the novel:

> the girls and the men were very dark in colour, but not more so than the South Italians or Spaniards. The men wore no trousers, but were dressed nearly the same as the Arabs whom I have seen in Algeria. They were of the most magnificent presence, being no less strong and handsome than the women were beautiful; and not only this, but their expression was courteous and benign.[7]

The language of this foreign land is also similar, in both tone and accent, to Italian. Even the boat that at the end of the novel transports the protagonist back to Europe is Italian: 'the Principe Umberto, bound from Callao to Genoa'.[8] In other words, *Erewhon* is a tribute to Italy.

When it comes to the reception of Butler in Italy, it is easy to see that Italians, from the north to the south, always demonstrated a great deal of respect for the British writer. In contrast to his negative reputation in Britain, in Italy Butler's literary and scientific works were perceived in a totally different light. This different reception was due to several factors that can help us to shed some additional light on the reasons why Butler was so respected in Italy during his lifetime and the posthumous fame of this Victorian polymath in wider Europe.

The Reception of Butler in Italy and France

Before his death, Butler was known and respected for his novels, artistic works, and even for his bizarre scientific ideas in Italy. He was an appreciated literary man in Sicily, and a recognized artist, photographer, and traveller in Piedmont and Lombardy. When we look at what Elinor Shaffer's *Erewhons of the Eye* tells us about the bond between Butler and Italy, it appears immediately clear how, for the Victorian writer, this southern European country was more than a second home. Indeed, in her critical account of Butler's artistic career, Shaffer distinguished two prominent places: London as the classroom and Italy as the training ground.[9]

Butler, of course, confirms his love for the country in most of his books. For instance, at the beginning of *Alps and Sanctuaries of Piedmont and the Canton Ticino*, Butler explains how Italy was a place for inspiration also shared by other illustrious artists like Handel and Shakespeare:

> It is always a pleasure to me to reflect that the countries dearest to these two master spirits are those which are also dearest to myself, I mean England and Italy. Both of them lived mainly here in London, but both of them turned mainly to Italy when realising their dreams. Handel's music is the embodiment of all the best Italian music of his time and before him, assimilated and

reproduced with the enlargements and additions suggested by his own genius. He studied in Italy; his subjects for many years were almost exclusively from Italian sources; the very language of his thoughts was Italian, and to the end of his life he would have composed nothing but Italian operas, if the English public would have supported him. His spirit flew to Italy, but his home was London. So also Shakespeare turned to Italy more than to any other country for his subjects. Roughly, he wrote nineteen Italian, or what to him were virtually Italian plays, to twelve English, one Scotch, one Danish, three French, and two early British.[10]

Published in 1881, *Alps and Sanctuaries* was the first book written by Butler about Italian art and culture. The book, a sort of travel guide, is a report of Butler's journeys across Piedmont and Lombardy. Partially a guide about the art and partially an ethnographic description of the local population and traditions, *Alps and Sanctuaries* is Butler's main tribute to Italy. The strong partnership with and love for the country was further enhanced by Butler's second book on Italian art: *Ex Voto*. The volume, a gift from Butler to the people of Varallo Sesia (a town in northern Piedmont) is a study and commentary of the Sacro Monte di Varallo ('Sacred Mountain of Varallo') and the art of Jan de Wespin Tabachetti (1568–1615) a Flemish architect and artist working at Crea. *Ex Voto* is dedicated '[a]i Varallesi e Valsesiani l'autore riconoscente' ('to the people of Varallo and Valsesia the grateful author'). As noted by Attilio Sella, an Italian journalist friend of Butler, the writer, after a dinner with the locals at the Sacro Monte, promised to write a book dedicated to the art of that wonderful town.[11]

The cultural and social bonds between Butler and Italy were not limited to the north of the country. Since the 1890s Butler also visited the south, in particular Sicily. Butler's travels to the south of the country were mostly due to his new controversial research on the *Odyssey*. This aspect of Butler's work has been studied in detail, in recent years, especially by Renato Lo Schiavo.[12] Butler's project aimed at demonstrating that the *Odyssey* was written by a Sicilian woman called Princess Nausicaa. A note in his *Notebooks* sheds some light on this peculiar theory:

> The finding out that the *Odyssey* was written at Trapani, the clearing up of the whole topography of the poem, and the demonstration, as it seems to me, that the poem was written by a woman and not by a man. Indeed, I may almost claim to have discovered the *Odyssey*, so altered does it become when my views of it are adopted. And robbing Homer of the *Odyssey* has rendered the *Iliad* far more intelligible; besides, I have set the example of how he should be approached.[13]

In Butler's hypothesis most of the adventures of Odysseus, the protagonist, are set around Sicily and consequently, the author of the poem (or rather the authoress) could only have hailed from the town of Trapani.

Butler considered his work on the *Odyssey* as the awakening of an old young princess. Indeed, Butler — in the note entitled 'Nausicaa and Myself' — explained with his traditional sarcastic tone the importance of this research:

> I am elderly, grey-bearded and, according to my clerk, Alfred, disgustingly fat; I wear spectacles and get more and more bronchitic as I grow older. Still no

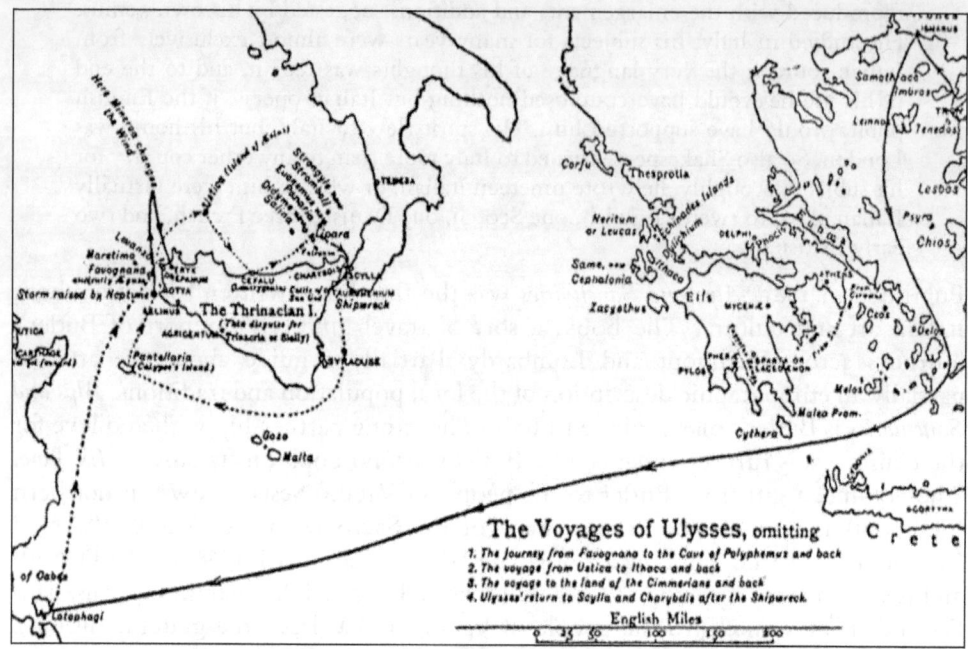

Figure: 'The Voyage of Ulysses', originally published in Samuel Butler, *The Authoress of the Odyssey* (New York: Dutton & Co., 1922), p. 181.

young prince in a fairy story ever found an invisible princess more effectually hidden behind a hedge of dullness or more fast asleep than Nausicaa was when I woke her and hailed her as Authoress of the *Odyssey*. And there was no difficulty about it either — all one had to do was to go up to the front door and ring the bell.[14]

Besides the controversial content of Butler's new project, the research on the *Odyssey* allowed the writer to make new acquaintances among the locals. The Victorian thinker became good friends with several Sicilians, especially from Trapani and Erice. As a sign of respect for this friendly British polymath, the main journals of the city, such as *Il Lambrischini* and *Quo Vadis*, decided to appoint Butler as their foreign correspondent. Between 1890 and 1902 Butler published, in both English and Italian, several articles concerning his ideas regarding the classic poem.

As reported by *La Falce*, a local journal, Butler became so famous in the area that when he arrived in Trapani on 8 May 1898 for one of his annual visits to the region, the main personalities of the town all gathered at the station to welcome him.[15] Signs of appreciation came from all over Sicily, on 20 March 1893, Butler was awarded, by the Accademia di scienze, lettere e belle arti of Acireale, the title of correspondent by the same institution.[16] The Accademia was, at the time, one of the most ancient academies in Italy. Opened in 1671 in Acireale (a small Sicilian town in the east of the island), it provided economic support for researchers working in the fields of natural history, literature, and poetry. It was also entitled to give certificates *ad honorem* to scholars with an established national or international reputation. Instead of being ignored and even ridiculed as he was in England, in Italy Butler was considered a key intellectual.

As explained the beginning of this epilogue, Italy was not the only country in which Butler's ideas were received with interest. In France, Butler's work started to be discussed thanks to the work of translation done by the writer, translator, and literary critic Valery Larbaud (1881–1957).[17] In the early 1920s, Larbaud published a series of articles about Butler in French periodicals such as *La Nouvelle revue française* and *La Revue de Paris*. Without going into detail here, it should be noted that Larbaud made two interesting points concerning the work of this problematic British writer. Firstly, he recognized the tendency of the Victorians to obscure Butler's fame. In particular, Larbaud described Butler's reception in England as 'une conspiration du silence' ('a conspiracy of silence'), which really undermined the importance of his ideas.[18] Secondly, the French translator found in Butler a free and independent thinker able to popularize scientific theories and literary ideas against the predominant culture of his generation.[19]

Along the same lines of Larbaud's article, in August 1921, the *Revue des deux mondes* published an article on Butler in the section 'Littératures étrangères', signed by the literary and art historian Louis Gillet (1876–1943).[20] The article, entitled 'Le Renommée posthume de Samuel Butler', defined Butler as one of those figures that needed to be rediscovered by the new European generations of the early twentieth century. Gillet, in introducing Butler's ideas to the French general public, discussed his whole work including his scientific ideas. In particular, Gillet framed *Life and Habit* and *Evolution, Old and New* in the new light of the early twentieth-century biological debate. The French historian even considered thinking about 'The Book of the Machines', the controversial section about the evolution of technology in the novel *Erewhon*, as a possible example of scientific writing.[21]

Therefore it is no surprise to find, in 1934, in the French periodical *Europe* another positive article dedicated to the English writer and entitled 'Les carnets de Samuel Butler'.[22] In the article Butler is described as a sort of cultural hero.[23] A hero who championed the tradition of his time like French philosophers did during the Enlightenment. This article is such an interesting example that shows how, for French intellectuals, Butler's critical attitude toward science and literature was perfectly acceptable even if he was not a professional.

The Reception of Butler's Science

While Butler's scientific ideas made little headway in England, they fared better abroad. In the early twentieth century, Butler's science of the mind became a topic of discussion among scientists and philosophers with an interest in neo-Lamarckism. Butler's work was, as explained by Forsdyke, re-evaluated and recognized in the twentieth century.[24] In particular, Forsdyke explains that 'we can now trace the path of fundamental informational ideas — oscillating between German and English — from Hering/Butler to Semon (1904), to Schrödinger, and then on to Francis Crick and others'.[25] This re-evaluation of Butler's science of the mind was possible thanks to the work of two key figures, the English professor of natural history Marcus Hartog and the Italian engineer and philosopher Eugenio Rignano.

Marcus Hartog is central to the new 'understanding' of Butler's science in the early twentieth century. Educated in biology and an expert in natural history, Hartog was one of the major followers of Butler's theory of memory and heredity. His interpretation of Butler's work focused on the assumption that Butler's notion of 'unconscious memory' was not an isolated case in Europe. In his introduction of the 1910 edition of *Unconscious Memory*, Hartog pointed out how Butler's evolutionary idea was significant, especially when linked to the work of Hering and Ribot. For Hartog it was necessary to think about the work of Butler within a wider European context where similar ideas where discussed. He wrote: '*Unconscious Memory* was largely written to show the relation of Butler's views to Hering's, and contains an exquisitely written translation of the Address.'[26] Nonetheless, Hartog also recognized continuity between their conceptualization of memory that went beyond the evident similarity in theories.

In 1914 Hartog published, in the Italian periodical *Scientia*, the article, 'Samuel Butler and Recent Mnemic Biological Theories', where he defined Butler as one of the most unique spirits of the whole Victorian period.[27] He explained that Butler's finest merit was not only that of being able to popularize evolution to a general audience but also being an inspiration to science in the twentieth century.[28] Hartog made clear in his article that:

> The name of Samuel Butler is one that has been growing more and more familiar to English biologists of late years: it is little, if at all to others, save for quotations in English writers, and a couple of references in R. Semon in his *Die Mneme*.[29]

Interestingly, Hartog recognized the growing importance of Butler's theory of memory and heredity among the professional community. In particular, the English biologist insisted in explaining the importance of thinking about the theory and not about the controversies included in Butler's work.

In 1913, Hartog published another book that relied, although only partially, on Butler's theory of memory and heredity: *Problems of Life and Reproduction*. Clearly an example of neo-Lamarckism, the volume contained a long and exhaustive discussion of Butler's work where Hartog recognized the significance of his science of the mind. In this book, Hartog referred to an anecdote which can help us to understand how and where Butler's ideas started to be taken more seriously. Hartog mentioned a talk given by Francis Darwin and Alfred Russell Wallace, during the Darwin celebrations of 1908–09. In their talk, Darwin and Wallace acknowledged Butler's name among other neo-Lamarckian scientists and philosophers. Hartog wrote:

> In this address we find the theory of Hering, Butler, Rignano, and Semon taking its proper place as a *vera causa* of that variation which Natural Selection must find before it can act, and recognised as the basis of a rational theory of the development of the individual and of the race.[30]

In *Problems of Life and Reproduction*, Hartog also dedicated some words to Butler's writing style and non-professional status. The zoologist explained that books like *Life and Habit* and *Unconscious Memory* were written with

> singular freshness and an ingenuity which compensates for the author's avowed lack of biological knowledge. This theory has indeed a tentative character, and lacks symmetrical completeness, but is the more welcome as not aiming at the impossible. A whole series of phenomena in organic beings are correlated under the term of memory, *conscious* and *unconscious*, patent and latent.[31]

It is no surprise, then, to see how Butler's work started to be considered as the work of a pioneer in the study of biological memory, but only when the reader put aside Butler's peculiar writing style and his personal feelings toward other scientists.

Hartog published most of his articles on Butler in the Italian journal *Scientia*. It is important to highlight that *Scientia*, at the beginning of the twentieth century, was publishing the 'avant-garde' of scientific and philosophical articles about evolution, psychology, and the mind. The history of this periodical is particularly important because it helps us to understand why there was a return to Butler's work in that particular context. *Scientia* was created with the aim of providing a solution to the loss of knowledge produced by the specialization in science which, in the opinion of the journal's editors, did not allow a proper exploration of any scientific or medical topic. The editorial committee was very specific about this issue. They stressed the need to go beyond the limits imposed by specialisms and to think about and explore key scientific questions using a more comprehensive approach which also involved philosophical questions.[32] This attitude was confirmed by the editorial committee of the journal which included: the Professor of Mathematics at the University of Bologna Federigo Enriques (1871–1946), the chemist Giuseppe Bruni (1873–1946) from the University of Padua, the medical doctor Antonio Dionisi (1866–1931) from Modena, Andrea Giardina (1875–1948) — biologist and zoologist from Palermo, and, finally, one of the main neo-Lamarckian thinkers in Italy, the engineer Eugenio Rignano from Milan.

Among these Italian scientists, Rignano was particularly important for the revival of Butler's contribution to the scientific debate. He published widely on philosophical and scientific topics. Educated at the University of Pisa and then at the Polytechnic of Turin, Rignano is a significant figure in the reception of Butler's work for two main reasons. First, Rignano recognized a continuity between the work of the Victorian writer and that of Hering. Second, and more importantly, Rignano promoted a reading of Hering's and Butler's theory of memory and heredity as a key component of the new neo-Lamarckian biological understanding of the hereditary process which would be widely discussed in the early twentieth century.

In his work, Rignano was concerned with the debate between neo-Lamarckians and neo-Darwinians relative to the transmissibility or non-transmissibility of acquired characters. His main book, *Sulla trasmissibilità dei caratteri acquisiti* (1906), explained the process of inheritance in a manner very similar to that of Hering (and indirectly Butler). Rignano, in particular, claimed that '[t]he comparison between the phenomena of development and the phenomena of memory especially after the discovery of the fundamental biogenetic law, that ontogeny is a recapitulation of phylogeny, has presented itself spontaneously to a large number of authors.'[33] These authors include Ewald Hering, Théodule-Armand Ribot, Claude Bernard and, although not mentioned directly, Butler. Without wanting to go too much into detail here, it is important to note that the theory of memory and heredity was considered as a fundamental part of the new biological debate concerning the transmission of acquired characteristics.[34]

The writings of both Hartog and Rignano present an unexpected portrait of Butler's idea of memory and heredity. During his lifetime, Butler's work was neglected and ignored by the British scientific community to whom he tried to communicate it. However, Butler was posthumously recognized not only as a popularizer of Hering's ideas, but also a valuable contributor to the debate about the mechanisms of evolution. Rignano was not alone in placing Butler's name alongside that of Ribot and Hering as an important figure. At the beginning of the twentieth century, Butler was instead considered a relevant figure of the post-Darwinian debate on heredity. This is such an important shift in the reception of Butler's theory. For the first time, we can see how his ideas started to be taken seriously among biologists in the early twentieth century.

In 1923, the Lithuanian scientist Sergei J. Tomkeieff published an article in *Scientia* entitled 'The Mnemic Theories of Evolution'. In the article, he provided an overview of the works and ideas of various figures interested in the relationship between memory and heredity in evolution such as Hering, Butler, Cope, and Rignano. Specifically writing about Butler, Tomkeieff defined the Victorian as the author 'whose genius is not yet fully appreciated even in his own country'.[35] Tomkeieff recognized the importance of Butler's ideas and even praised his ability to come to such important conclusions without any scientific training.[36] Tomkeieff, in addition, pointed out a strong correlation between the Butler-Hering theory and the work of the American palaeontologist and biologist Edward Cope (1840–1897). In his work, Cope deployed a neo-Lamarckian understanding of evolution which presented a theory of memory almost identical to the one advanced by Butler.[37] Indeed, as Tomkeieff observed, 'many of the passages in Cope's works remind us of the vivid passages of Butler.'[38] Beside the specific similarity between the two theories of memory and heredity examined, what appears clear is that the significance of Butler's work was finally recognized and taken into account by professional scientists.

Having said that, the endeavour of Hartog, Rignano, and Tomkeieff remained an isolated event in the history of twentieth-century biology. By the end of the 1920s, Butler's work on evolution and the mind was long forgotten, and this controversial Victorian polymath returned to being just the divisive author of *Erewhon* and *The Way of all Flesh*. I would like to conclude this volume with the hope that this work has, at least, shed some light upon the place of Butler in the late nineteenth-century pan-European debate on evolution, heredity, and the mind.

Notes to the Conclusion

1. Butler, *The Notebooks*, p. 165.
2. Stanley Bates Harkness, *The Career of Samuel Butler, 1835–1902: A Bibliography* (London: Bodley Head, 1955), p. 15.
3. Harkness, *The Career of Samuel Butler*, pp. 14–16.
4. Recent studies have discussed the role played by the Italian peninsula in the Victorian imagination exploring, in particular, the complex relationship between English and Italian literature and the influence that British culture and science had on the Italian Risorgimento. For instance, Victorian writers and poets as diverse as Christina Rossetti (1830–1894), Charles Dickens, Robert Browning (1812–1889), Coventry Patmore (1823–1896), and George Eliot experienced, lived, and transplanted into their writings some aspect of Italian culture and art. See *The Victorians and Italy: Literature, Travel, Politics and Art*, ed. by Alessandro Vescovi, Luisa Villa, and Paul Vita, (Monza: Polimetrica, 2007); Hilary Fraser, *The Victorians and Renaissance Italy* (London: Wiley-Blackwell, 1992); John Easton Law and Lene Østermark-Johansen, *Victorian and Edwardian Responses to the Italian Renaissance* (London: Routledge, 2005).
5. See Shaffer, *Erewhons of the Eye*, pp. 67–167.
6. Jones, Samuel Butler, *Samuel Butler, Author of Erewhon (1835–1902)*, I, 26–28.
7. Butler, *Erewhon*, p. 49.
8. Butler, *Erewhon*, p. 314.
9. Shaffer's book provides an analysis of many of the photographs Butler took over several journeys through the country. Those images, as rightly observed by Shaffer, demonstrate that the British writer was deeply in love with Italy and its people. See, Shaffer, *Erewhons of the Eye*, pp. 17–20.
10. Butler, *Alps and Sanctuaries*, pp. 18–19.
11. Attilio Sella wrote a series of articles which were published between 1898 and 1900 in the *Gazzetta di Novara*; one of them is important here: see Attilio Sella, ʻUnʼ Inglese fervido amico dell'Italia, Samuel Butler' (1916), a copy of which was given by Sella to Henry Festing Jones. See St Johns College Library, Cambridge, Samuel Butler Collection, Series Section 3 (III): Books, etc., about Samuel Butler — Butler/Section 3 (III)/7/7.
12. See Renato Lo Schiavo, *La teoria dell'origine siciliana dell'Odissea: Il cieco, la giovinetta ed il malconsiglio* (Palermo: ISSPE, 2003).
13. Butler, *The Notebooks*, p. 376.
14. Butler, *The Notebooks*, p. 193.
15. Extract from *La Falce*, 1.20, 15 May 1898, [n.p]. bibliography. The passage is cited in Giovanni Angelo, 'Samuel Butler in Sicilia', *Il Baretti*, 43/44, (1967), 76–91 (p. 79).
16. The original certificates are kept in the Butler Collection, St John's College Library, Cambridge: Butler/Section 8 (VIII)/2/3/1; Butler/Section 8 (VIII)/2/3/2.
17. Apart from his translation of Butler's work, Larbaud also translated and popularized the works of Samuel Taylor Coleridge, Walt Whitman, and James Joyce, whose *Ulysses* was translated by Auguste Morel (1924–29) under Larbaud's supervision. Larbaud's translation of Butler's work is particularly important here as it shows that Butler's ideas also had significance among European intellectuals.
18. Valery Larbaud, 'Samuel Butler', *La Nouvelle revue française*, 76 (1920), 5–37 (p. 6).
19. Valery Larbaud, 'Samuel Butler', *La Revue de Paris*, 30.16 (1923), 748–62 (p. 749).
20. Louis Gillet, 'Littératures étrangères: La renommée posthume de Samuel Butler', *Revue des deux Mondes*, 64.3 (1921), 683–96.
21. Gillet, 'Littératures étrangères, pp. 690–91.
22. Extracts from Butler's notebooks were published in Samuel Butler, 'Les carnets de Samuel Butler', *Europe*, 144.1 (15 December 1934), 485–505.
23. Among other things, Butler is described as an individual whose eclectic work can be compared to that of the French encyclopaedists of the late eighteenth century and who even — had he lived a century before — would have been, beyond any doubt, a good friend of Voltaire and Chamfort. Here is the original in French: 'avec Samuel Butler nous retrouvons l'ordre avec l'abondance', and '[s]on éclectisme, la diversité de ses intentions rappellent nos encyclopédistes,

son style prolonge le rapprochement. Un siècle plus tôt il eut été sans doute le grand ami de Voltaire et de Chamfort.' See Butler, 'Les carnets des Samuel Butler', p. 485.
24. Donald R. Forsdyke, '"A Vehicle of Symbols and Nothing More": George Romanes, Theory of Mind, Information, and Samuel Butler', *History of Psychiatry*, 26.3 (2015), 270–87 (pp. 276–78).
25. Forsdyke, '"A Vehicle of Symbols and Nothing More"', 270–87 (p. 276).
26. Butler, *Luck or Cunning?*, pp. 15–16.
27. Marcus Hartog, 'Samuel Butler and Recent Mnemic Biological Theories', *Scientia*, 15.33 (1914), 37–52 (p. 40).
28. Hartog, 'Samuel Butler and Recent Mnemic Biological Theories', p. 55.
29. Hartog, 'Samuel Butler and Recent Mnemic Biological Theories', p. 1.
30. Marcus Hartog, *Problems of Life and Reproduction* (London: Murray, 1913), p. 280.
31. Hartog, 'Samuel Butler and Recent Mnemic Biological Theories', p. 77.
32. See Editorial committee, 'Programma', *Rivista di Scienza (Scientia)*, 1.1 (1907), 1–3 (p. 1).
33. Eugenio Rignano, *Upon the Inheritance of Acquired Characters: A Hypothesis of Heredity, Development, and Assimilation* (Chicago: The Open Court Publishing Co., 1911), p. 316.
34. What is particularly striking in Rignano's work is how he insisted on the mnemonic process as a concrete possibility in explaining the hereditary mechanism. In his review of August Pauly's *Darwinismus und Lamarckismus* (1905), Rignano explained how the position of Pauly had developed — starting from the big discoveries in the organic memory debate advanced by neo-Lamarckian thinking like Hering's. See Eugenio Rignano, review of August Pauly, *Darwinismus und Lamarckismus*, in *Scientia*, 1 (1907), 192–98 (p. 195).
35. S. J. Tomkeieff, 'The Mnemic Theories of Evolution', *Scientia*, 34 (1923), 159–72 (p. 160).
36. Tomkeieff, 'The Mnemic Theories of Evolution', p. 160.
37. Tomkeieff, 'The Mnemic Theories of Evolution', p. 162. In addition, regarding Cope's work on evolution see the recent David Ceccarelli, 'Between Social and Biological Heredity: Cope and Baldwin on Evolution, Inheritance, and Mind', *Journal of the History of Biology*, 52.1 (2019), 161–94.
38. Tomkeieff, 'The Mnemic Theories of Evolution', p. 163.

BIBLIOGRAPHY

From Samuel Butler's archives

Library holdings

Cambridge, Cambridge University Library, Department of Manuscripts and University Archives
> Add 4251/191–202: letters mainly to Henry Festing Jones
> Add 5977: MS of *Unconscious Memory*
> MS DAR 92 & 106–07: letters to Charles Darwin and related papers

Cambridge, St John's College Library
> Samuel Butler Collection: papers of Samuel Butler; including books, atlases, and music from his collection, and prints, photographs, and portraits of and by him (c. 100 boxes, 1700 glass negatives, 600 printed volumes, 450 pictures, and 50 artefacts)

Christchurch, Canterbury Museum, New Zealand
> ZB823BUT: letter book and sketchbook

London, British Library, Manuscript Collections
> Add MSS 36711–13, 38176–77, 39846–47, 44027–54: correspondence and papers including literary MSS
> Add MSS 44027–44042: general correspondence, including family and business letters
> Add MSS 44028–44042: correspondence with Henry Festing Jones
> Add MS 44043: correspondence with Eliza Mary Anne Savage
> Add MSS 44045–44050: notebooks (copy B, the first pressed (autograph) copy)
> Add MS 71695: notebook (containing extracts from Herbert-Spencer's *Principles of Psychology*)

Oxford, Bodleian Library, Special Collections and Western Manuscripts
> MS Eng lett c 782, fols 84–87: letters
> MS Eng misc d 96: typescript of *Luck or Cunning*

Williamstown, Williams College, Chapin Library of Rare Books
> Chapin Library Butler Collection: correspondence and literary papers

Online sources

Darwin Correspondence Project <www.darwinproject.ac.uk>
Science in the Nineteenth-Century Periodical <http://www.sciper.org/>

General bibliography

ALLAMAN, GEORGE J., 'Protoplasm and Life', *Popular Science Monthly*, 15 (1879), 721–22
ALLEN, GRANT, REVIEW OF SAMUEL BUTLER, *Evolution, Old and New*, in *The Examiner* (17 May 1879), 646–47
——, REVIEW OF BUTLER, *Evolution, Old and New*, in *Academy*, 15 (1879), 426
ANGELO, GIOVANNI, 'Samuel Butler in Sicilia', *Il Baretti*, 43/44, (1967), 76–91
BARREAU, HERVÉ, 'Bergson face à Spencer: Vers un nouveau positivisme', *Archives de philosophie*, 71.2 (2008), 219–43
BARSANTI, GIULIO, *Una lunga pazienza cieca: Storia dell'evoluzionismo* (Torino: Einaudi, 2005)
BARTON, RUTH '"Huxley, Lubbock, and Half a Dozen Others": Professionals and Gentlemen in the Formation of the X Club, 1851–1864', *Isis*, 89.3 (1998), 410–44
——'"An Influential Set of Chaps": The X-Club and Royal Society Politics 1864–85", *The British Journal for the History of Science*, 23.1 (1990), 53–81
BAUMANN, C., 'Ewald Hering's Opponent Colors: History of an Idea', *Der Ophthalmologe: Zeitschrift der Deutschen Ophthalmologischen Gesellschaft*, 89.3 (1992), 249–52
BEER, GILLIAN, 'Butler, Memory, and the Future', in *Samuel Butler, Victorian against the Grain: A Critical Overview*, ed. by James Paradis (Toronto: University of Toronto Press, 2007), pp. 45–57
——'Darwin and Romanticism', *The Wordsworth Circle*, 41.1 (2010), 3–9
——*Darwin's Plots: Evolutionary Narrative in Darwin, George Eliot and Nineteenth-Century Fiction* (Cambridge: Cambridge University Press, 2000)
——*Open Fields: Science in Cultural Encounter* (Oxford: Clarendon Press, 1996)
BERGSON, HENRI, *L'Évolution créatrice* (Paris: Félix Alcan, 1907)
BERNARD, CLAUDE, *An Introduction to the Study of Experimental Medicine*, trans. by H. C. Greene (New York: Schuman, 1949)
BIGONI, FRANCESCA, and GIULIO BARSANTI, 'Evolutionary Trees and the Rise of Modern Primatology: The Forgotten Contribution of St. George Mivart', *Journal of Anthropological Sciences*, 89 (2011), 93–107
BOWLER, PETER, *The Eclipse of Darwinism: Anti-Darwinian Evolution Theories in the Decades around 1900* (Baltimore: Johns Hopkins University Press, 1992)
——*Evolution: The History of an Idea* (Berkley: University of California Press, 1989)
BROWNE, JANET, *Charles Darwin: The Power of Place* (London: Cape, 2002)
BURROWS, MARK S., 'A Historical Reconsideration of Newman and Liberalism: Newman and Mivart on Science and the Church', *Scottish Journal of Theology*, 40.3 (1987), 399–419
BUTLER, SAMUEL, *Alps and Sanctuaries of Piedmont and the Canton Ticino* (London: Bogue, 1882)
——*The Authoress of the Odyssey* (New York: Dutton & Co., 1922)
——'Les carnets de Samuel Butler', *Europe*, 144.1 (15 December 1934), 485–505
——'Darwin among the Machines', in *A First Year in Canterbury Settlement: With Other Early Essays* (London: Fifield, 1914), pp. 179–85; also available in *The Notebooks of Samuel Butler*, ed. by Henry Festing Jones (London: Fifield, 1912), pp. 42–47
——'The Deadlock of Darwinism', in Samuel Butler, *Essays on Life, Art, and Science* (London: Fifield, 1908), pp. 234–340
——*Erewhon, or, Over the Range* (London: Fifield, 1908)
[ANON.], REVIEW OF SAMUEL BUTLER, *Erewhon: or, Over the Range*, in *The Athenaeum*, 2321 (20 April 1872), 492
[ANON.], REVIEW OF BUTLER, *Erewhon [...]*, in *British Quarterly Review*, 56 (July 1872), 261–63
[ANON.], REVIEW OF BUTLER, *Erewhon [...]*, in *Fortnightly Review*, 11.65 (May 1872), 609–10
BUTLER, SAMUEL, *Essays on Life, Art and Science* (London: Fifield, 1908)
——*The Evidence of the Resurrection of Christ* (London: [n. pub.], 1865)

—— *Evolution, Old and New, or, The Theories of Buffon, Dr. Erasmus Darwin and Lamarck, as Compared with that of Charles Darwin* (London: Fifield, 1911)
[ANON.], REVIEW OF SAMUEL BUTLER, *Evolution, Old and New*, in *The Examiner* (17 May 1879), 646–47
[ANON.], REVIEW OF BUTLER, *Evolution, Old and New*, in *Saturday Review* (31 May 1879), 682–84
BUTLER, SAMUEL, *Ex Voto: An Account of the Sacro Monte or New Jerusalem at Varallo-Sesia* (London: Trübner and Co., 1888)
—— *The Family Letters of Samuel Butler, 1841–1866*, ed. by Arnold Silver (Palo Alto: Stanford University Press, 1962)
—— *A First Year in Canterbury Settlement: With Other Early* Essays (London: Fifield, 1914)
—— *God the Known and God the Unknown* (New Haven: Yale University Press, 1917)
—— *Letters between Samuel Butler and Miss E. M. A. Savage, 1871–1885* (London: Cape, 1935)
—— *Life and Habit* (London: Cape, 1910)
[ANON.], REVIEW OF SAMUEL BUTLER, *Life and Habit*, in *Daily News* (20 January 1880), [n.p.]
[ANON.], REVIEW OF BUTLER, *Life and Habit*, in *The Saturday Review* (26 December 1878), 119–21
BUTLER, SAMUEL, *Luck, or Cunning, as the Main Means of Organic Modification? An Attempt to Throw Additional Light upon Charles Darwin's Theory of Natural Selection* (London: Fifield, 1910)
—— 'Lucubratio Ebria', in *A First Year in Canterbury Settlement: With Other Early Essays* (London: Fifield, 1914), pp. 186–94; also available in *The Notebooks of Samuel Butler*, ed. by Henry Festing Jones (London: Fifield, 1912), pp. 47–53
—— *The Notebooks of Samuel Butler*, ed. by Henry Festing Jones (London: Fifield, 1912)
—— *Selections from Previous Works and Remarks on Romanes' Mental Evolution in Animals* (London: Trübner & Co., 1884)
—— *Unconscious Memory: A Comparison Between the Theory of Dr. Ewald Hering, Professor of Physiology in the University of Prague, and the 'Philosophy of the Unconscious' of Dr. Edward von Hartmann, with Translations from both these Authors, and Preliminary Chapters Bearing upon 'Life and Habit', 'Evolution, Old and New', and Mr. Charles Darwin's Edition of Dr. Krause's 'Erasmus Darwin'* (London, Cape: 1920)
[ANON.], REVIEW OF SAMUEL BUTLER, *Unconscious Memory: A Comparison Between the Theory of Dr. Ewald Hering, Professor of Physiology in the University of Prague, and the 'Philosophy of the Unconscious' of Dr. Edward von Hartmann, with Translations from both these Authors, and Preliminary Chapters Bearing upon 'Life and Habit', 'Evolution, Old and New', and Mr. Charles Darwin's Edition of Dr. Krause's 'Erasmus Darwin'*, in *The Athenaeum*, 2773 (18 December 1880), 810
[ANON.], REVIEW OF BUTLER, *Unconscious Memory [...]*, in *The British Journal of Homeopathy*, 155 (1880), 67–73
[ANON.], REVIEW OF BUTLER, *Unconscious Memory [...]*, in *St. James's Gazette*, 2 December 1880, 13
[ANON.], REVIEW OF BUTLER, *Unconscious Memory [...]*, in *The Journal of Science*, 3 (1881), 44
BUTLER, SAMUEL, *The Way of all Flesh* (New York: Dutton & Co., 1916)
CANTOR, GEOFFREY, and OTHERS, *Science in the Nineteenth-Century Periodical: Reading the Magazine of Nature* (Cambridge: Cambridge University Press, 2008)
The Cambridge Companion to Darwin, ed. by Jonathan Hodge and Gregory Radick (Cambridge: Cambridge University Press, 2003)
The Cambridge Companion to the 'Origin of Species', ed. by Michael Ruse and Robert J. Richards (Cambridge: Cambridge University Press 2009)
The Cambridge History of English and American Literature: An Encyclopaedia in Eighteen Volumes,

ed. by A. W. Ward and others, 18 vols (Cambridge: Cambridge University Press, 1907–2000), XII: *The Romantic Revival*, ed. by A. W. Ward and A. R. Waller (1915)

CARPENTER, WILLIAM BENJAMIN, *Principles of Mental Physiology* (London: King & Co., 1874)

CARROY, JACQUELINE, and OTHERS, 'Les entreprises intellectuelles de Théodule Ribot', *Revue philosophique de la France et de l'étranger*, 4 (2016), 451–64

CECCARELLI, DAVID, 'Between Social and Biological Heredity: Cope and Baldwin on Evolution, Inheritance, and Mind', *Journal of the History of Biology*, 52.1 (2019), 161–94

CHAMBERS, ROBERT, *Vestiges of the Natural History of Creation* (London: Churchill, 1844)

COHEN, PHILIP, 'Stamped on His Work: The Decline of Butler's Literary Reputation', *The Journal of the Midwest Modern Language Association*, 18.1 (1985), 65–66

COLERIDGE, SAMUEL TAYLOR, *Hints towards the Formation of a More Comprehensive Theory of Life* (London: John Churchill, 1818)

CORSI, PIETRO, *The Age of Lamarck: Evolutionary Theories in France, 1790–1830* (Berkeley: University of California Press, 1998)

DANZIGER, KURT, *Constructing the Subject: Historical Origins of Psychological Research* (Cambridge: Cambridge University Press, 1994)

—— 'The Positivist Repudiation of Wundt', *Journal of the History of the Behavioral Sciences*, 15 (1979), 205–30

DARWIN, CHARLES, *The Autobiography of Charles Darwin, 1809–1882: With the Original Omissions Restored*, ed. and with an appendix and notes by Nora Barlow (London: Collins, 1958)

—— *On the Origin of Species by Means of Natural Selection, or the Preservation of Favoured Races in the Struggle for Life*, 6th edn, with additions and corrections (London: Murray, 1872)

—— *The Variation of Animals and Plants under Domestication* (London: Murray, 1868)

DARWIN, FRANCIS, *Life and Letters of Charles Darwin* (London: Murray, 1887)

DE LANESSAN, JEAN-LOUIS, *Le Transformisme: Évolution de la matière et des êtres vivants* (Paris: Octave Doin, 1883)

DESMOND, ADRIAN J., and JAMES RICHARD MOORE, *Darwin* (London: Norton & Co., 1992)

DESMOND, ADRIAN J., *Archetypes and Ancestors: Palaeontology in Victorian London* (Chicago: University of Chicago Press, 1984)

—— *The Politics of Evolution: Morphology, Medicine, and Reform in Radical London* (Chicago: University of Chicago Press, 1989)

—— 'Redefining the X Axis: "Professionals", "Amateurs" and the Making of Mid-Victorian Biology: A Progress Report', *Journal of the History of Biology*, 34.1 (2001), 3–50

DE VRIES, HUGO, *Intracellular Pangenesis* (Chicago: Chicago University Press, 1910)

EDITORIAL COMMITTEE, 'Programma', *Rivista di Scienza (Scientia)*, 1.1 (1907), 1–3

FORSDYKE, DONALD R., 'Heredity as Transmission of Information: Butlerian Intelligent Design', *Centaurus*, 48 (2006), 133–48

—— *The Origin of Species, Revisited: A Victorian Who Anticipated Modern Developments in Darwin's Theory* (Montreal: McGill-Queen's University Press, 2001)

—— 'Samuel Butler and Human Long Term Memory: Is the Cupboard Bare?', *Journal of Theoretical Biology*, 258 (2009), 156–64

—— '"A Vehicle of Symbols and Nothing More": George Romanes, Theory of Mind, Information, and Samuel Butler', *History of Psychiatry*, 26.3 (2015), 270–87

FRASER, HILARY, *The Victorians and Renaissance Italy* (London: Wiley-Blackwell, 1992)

FURBANK, P. N., *Samuel Butler, 1835–1902* (Cambridge: Cambridge University Press, 1948)

FYFE, AILEEN, *Science and Salvation: Evangelical Popular Science Publishing in Victorian Britain* (Chicago: University of Chicago Press, 2004)

——, and BERNARD LIGHTMAN, 'Science in the Marketplace: An Introduction', in *Science in the Marketplace: Nineteenth-Century Sites and Experiences*, ed. by Aileen Fyfe and Bernard Lightman (Chicago: University of Chicago Press, 2007), pp. 1–19

GILLET, LOUIS, LITTÉRATURES ÉTRANGÈRES: LA RENOMMÉE POSTHUME DE SAMUEL BUTLER', *Revue des deux Mondes*, 64.3 (1921), 683–96
GILLOTT, DAVID, *Samuel Butler against the Professionals: Rethinking Lamarckism 1860–1900* (London: Legenda, 2015)
GLICK, THOMAS F., *What about Darwin? All Species of Opinion from Scientists, Sage, Friends and Enemies Who Met, Read and Discussed the Naturalist Who Changed the World* (Baltimore: Johns Hopkins University Press, 2010)
GOOCH, JOSHUA A., 'Figures of Nineteenth-Century Biopower in Samuel Butler's *Erewhon*', *Nineteenth-Century Contexts: An Interdisciplinary Journal*, 36.1 (2014), 53–71
GRENDON, FELIX, 'Samuel Butler's God', *The North American Review*, 208 (1918), 277–86
GULLIN, VINCENT, 'Théodule Ribot's Ambiguous Positivism: Philosophical and Epistemological Strategies in the Founding of French Scientific Psychology', *Journal of the History of the Behavioral Sciences*, 40 (2004), 165–81
HAAST, JULIUS, *Geology of the Provinces of Canterbury and Westland* (Christchurch: printed at the *Times* Office, 1879)
HAJDENKO-MARSHALL, C., 'Believing after Darwin: The Debates of the Metaphysical Society (1869–1880)', *Cahiers victoriens et édouardien online*, 76 (2012), 69–83
HARKNESS, STANLEY BATES, *The Career of Samuel Butler, 1835–1902: A Bibliography* (London: Bodley Head, 1955)
[ANON.], 'Hartmann's *Philosophy of the Unconscious*', review, in *The Modern Review: A Quarterly Magazine*, 5 (1884), 776–81
[ANON.], 'Von Hartmann's *Philosophy of the Unconscious*', review, in *The Spectator*, 23 (1884), 1111
HARTOG, MARCUS, *Problems of Life and Reproduction* (London: Murray, 1913)
—— 'Samuel Butler and Recent Mnemic Biological Theories', *Scientia*, 15.33 (1914), 37–52
HARVEY, E. NEWTON, 'Some Physical Properties of Protoplasm', *Journal of Applied Physics*, 9.2 (1938), 68–80
HARVEY, JOY, *Almost a Man of Genius: Clémence Royer, Feminism and Nineteenth-Century Science* (New Brunswick: Rutgers University Press, 1997)
HAWKINS, MIKE, *Social Darwinism in European and American Thought, 1860–1945: Nature as Model and Nature as Threat* (Cambridge: Cambridge University Press, 2008)
HERING, EWALD, *On Memory and the Specific Energies of the Nervous System* (Chicago: The Open Court Publishing Company, 1895)
[ANON.], REVIEW OF EWALD HERING, *On Memory and the Specific Energies of the Nervous System*, in *The Monist*, 6 (1895), 634
HERING, EWALD, 'On the Theory of Nerve-Activity', *The Monist*, 10.2 (1900), 167–87
HOFSTADTER, RICHARD, *Social Darwinism in American Thought* (Boston: Beacon Press, 1944)
HOLT, LEE, *Samuel Butler* (New York: Grosset & Dunlap, 1963)
—— 'Samuel Butler up to Date', *English Literature in Transition, 1880–1920*, 3.1 (1960), 17–21
HOLTERHOFF, K., 'The History and Reception of Charles Darwin's Hypothesis of Pangenesis', *Journal of the History of Biology*, 47 (2014), 661–95
HOWARD, DANIEL, *The Correspondence of Samuel Butler with his Sister May* (Berkeley: University of California Press, 1962)
HUXLEY, THOMAS H., *Darwiniana* (London: Appleton, 1893)
—— 'Mr Darwin's Critics', *Contemporary Review*, 18 (1871), 442–76
—— *On the Physical Basis of Life* (New Haven, Conn.: The College Courant, 1869)
—— *Scientific Memoirs: Selected from the Transactions of Foreign Academies of Science, and from Foreign Journals* (London: Taylor and Francis, 1853)
IRVINE, WILLIAM, *Apes, Angels, and Victorians: The Story of Darwin, Huxley, and Evolution* (New York: McGraw-Hill Book Company, 1955)

JABLONKA, EVA, and MARIOT J. LAMB, *Epigenetic Inheritance and Evolution: The Lamarckian Dimension* (Oxford: Oxford University Press, 1999)

JANKO, JAN, 'Mach and Hering's Physiology of the Senses', *Clio Medica*, 33 (1995), 89–96

JONES, HENRY FESTING, *Charles Darwin and Samuel Butler: A Step towards Reconciliation* (London: Fifield, 1911)

—— *Samuel Butler, Author of Erewhon (1835–1902): A Memoir*, 2 vols (London: Macmillan and Co. Limited, 1919)

—— *Samuel Butler: A Sketch* (London: Cape, 1913)

JONES, STEVE, *The Darwin Archipelago: The Naturalist's Career beyond Origin of Species* (London: Yale University Press, 2012)

JORDANOVA, LUDMILLA, *Lamarck* (Oxford: Oxford University Press, 1984)

KINGSLEY, CHARLES, 'The Charm of Birds', *Fraser's Magazine for Town and Country*, 75 (1867), 802–10

KOHN, DAVID, *The Darwinian Heritage* (Princeton: Princeton University Press, 2014)

KRAUSE, ERNST, *Erasmus Darwin*, trans. from the German by W. S. Dallas, with a preliminary notice by Charles Darwin (London: Murray, 1879)

LAMARCK, JEAN-BAPTISTE, *Histoire naturelle des animaux sans vertèbres* (Paris: Baillière, 1815)

—— *Philosophical Zoology* (London: MacMillan and Co., 1914)

LANARO, GIORGIO, *L'evoluzione, il progresso e la società industriale: Un profilo di Herbert Spencer* (Rome: La Nuova Italia Editrice, 1997)

LANKESTER, RAY E. 'Perigenesis v. Pangenesis: Haeckel's New Theory of Heredity', *Nature*, 14.350 (1876) 235–38

—— 'Samuel Butler', *La Nouvelle revue française*, 76 (1920), 5–37

—— 'Samuel Butler', *La Revue de Paris*, 30.16 (1923), 748–62

LAW, JOHN EASTON, and LENE ØSTERMARK-JOHANSEN, *Victorian and Edwardian Responses to the Italian Renaissance* (London: Routledge, 2005)

LEVINE, GEORGE, *Darwin and the Novelists: Patterns of Science in Victorian Fiction* (Chicago: University of Chicago Press, 1992)

—— *Darwin the Writer* (Oxford: Oxford University Press, 2011)

LEWES, GEORGE HENRY, *Problems of Life and Mind* (London: Trübner and Co., 1873)

LIGHTMAN, BERNARD, 'Afterword', in *The Routledge Research Companion to Nineteenth-Century British Literature and Science*, ed. by John Holmes and Sharon Ruston (Oxford: Routledge, 2017), pp. 438–41

—— 'A Conspiracy of One: Butler, Natural Theology, and Victorian Popularization', in *Samuel Butler, Victorian against the Grain: A Critical Overview*, ed. by James Paradis (Toronto: University of Toronto Press, 2007), pp. 113–43

—— *Victorian Popularizers of Science: Designing Nature for New Audiences* (Chicago: Chicago University Press, 2007)

—— '"The Voices of Nature": Popularising Victorian Science', in *Victorian Science in Context*, ed. by Bernard Lightman (Chicago: Chicago University Press, 1997), pp. 187–211

LO SCHIAVO, RENATO, *La teoria dell'origine siciliana dell'Odissea: Il cieco, la giovinetta ed il malconsiglio* (Palermo: ISSPE, 2003)

LYELL, CHARLES, *Geological Evidences of the Antiquity of Man* (London: Murray, 1863)

—— *Principles of Geology* (London: Murray, 1837)

LYTTON, EDWARD, *The Coming Race* (London: Blackwood and Sons, 1871)

MANDLER, GEORGE, *A History of Modern Experimental Psychology: From James and Wundt to Cognitive Science* (London: MIT Press, 2007)

MATTHEW, PATRICK, *On Naval Timber and Arboriculture: With Critical Notes on Authors Who Have Recently Treated the Subject of Planting* (London: Longman, Rees, Orme, Brown, and Green, 1831)

MIDDLETON, DARREN J. N., 'James Russell Perkin, Theology and the Victorian Novel', *Religious Studies Review*, 36.4 (2010), 289

MIVART, ST. GEORGE, 'Evolution and its Consequences: A Reply to Professor Huxley', *The Contemporary Review*, 19 (December 1871/May 1872), 168–97

—— *On the Genesis of Species* (London: Macmillan & Co., 1871)

MUCCHIELLI, LAURENT, 'Aux Origines de la psychologie universitaire en France (1870–1900): Enjeux intellectuels, contexte politique, réseaux et stratégies d'alliance autour de la *Revue philosophique* de Théodule Ribot', *Annals of Science*, 55 (1998), 263–89

MUGGERIDGE, MALCOLM, *The Earnest Atheist: A Study of Samuel Butler* (London: Eyre & Spottiswoode, 1936)

MYERS-SHAFFER, CHRISTINA, *The Principles of Literature: A Guide for Readers and Writers* (New York: Barron's Educational Series, 2000)

NEWMAN, JOHN HENRY, *The Idea of a University: Defined and Illustrated* (London: Pickering, 1873)

NICOLAS, SERGE, and AGNÈS CHARVILLAT, 'Introducing Psychology as an Academic Discipline in France: Théodule Ribot and the Collège de France (1888–1901)', *Journal of the History of the Behavioral Sciences*, 37 (2001), 143–64

O'CONNOR, RALPH, *The Earth on Show: Fossils and the Poetics of Popular Science, 1802–1856* (Chicago: The University of Chicago Press, 2007)

OTIS, LAURA, *Müller's Lab* (Oxford: Oxford University Press, 2007)

—— *Organic Memory: History and the Body in the Late Nineteenth and Early Twentieth Centuries* (Lincoln: University of Nebraska Press, 1994)

OTIS, LAURA, and S. NICOLAS, *Théodule Ribot: Philosophe breton, fondateur de la psychologie française* (Paris: L'Harmattan, 2005)

PARADIS, JAMES, 'Butler after Butler: The Man of Letters as Outsider', in *Samuel Butler, Victorian against the Grain: A Critical Overview*, ed. by James Paradis (Toronto: University of Toronto Press, 2007), pp. 343–70

—— 'The Butler-Darwin Biographical Controversy in the Victorian Periodical Press', in *Science Serialized: Representations of the Sciences in Nineteenth-Century Periodicals*, ed. by Geoffrey Cantor and Sally Shuttleworth (Cambridge: MIT Press, 2004), pp. 307–31

PAULY, PHILIP J., 'Samuel Butler and His Darwinian Critics', *Victorian Studies*, 25.2 (1982), 161–80

PERKIN, JAMES RUSSELL, *Theology and the Victorian Novel* (Montreal: McGill-Queen's University Press, 2009)

RABINBACH, ANSON, *The Human Motor: Energy, Fatigue, and the Origins of Modernity* (Oakland: University of California Press, 1992)

RABY, PETER, *Samuel Butler: A Biography* (London: Hogarth Press, 1991)

RACHELS, JAMES, *Created from Animals: The Moral Implications of Darwinism* (Oxford: Oxford University Press, 1991)

RADICK, GREGORY, *The Simian Tongue: The Long Debate about Animal Language* (Chicago: University of Chicago Press, 2007)

The Reception of Charles Darwin in Europe, ed. by Eve-Marie Engels and Thomas F. Glick (London: Bloomsbury, 2009)

RIBOT, THÉODULE, *L'Hérédité: Étude psychologique* (Paris: Baillière, 1873)

——, *Les Maladies de la mémoire* (Paris: L'Harmattan, 1881)

—— *La Psychologie anglaise contemporaine* (Paris: Ladrange, 1881)

RICHARDS, ROBERT, *Darwin and the Emergence of Evolutionary Theories of Mind and Behaviour* (Chicago: University of Chicago Press, 1989)

—— *The Meaning of Evolution: The Morphological Construction and Ideological Reconstruction of Darwin's Theory* (Chicago: University of Chicago Press, 1993)

——— *The Romantic Conception of Life: Science and Philosophy in the Age of Goethe* (Chicago: University of Chicago Press, 2002)

RIGNANO, EUGENIO, REVIEW OF AUGUST PAULY, *Darwinismus und Lamarckismus*, in *Scientia*, 1 (1907), 192–98

——— *Sulla trasmissibilità dei caratteri acquisiti: Ipotesi d'una centro-epigenesi* (Bologna: Zanichelli, 1906)

——— *Upon the Inheritance of Acquired Characters: A Hypothesis of Heredity, Development, and Assimilation* (Chicago: The Open Court Publishing Co., 1911)

ROBINSON, ROGER, 'From Canterbury Settlement to Erewhon: Butler and Antipodean Counterpoint', in *Samuel Butler, Victorian against the Grain: A Critical Overview*, ed. by James Paradis (Toronto: University of Toronto Press, 2007), pp. 21–44

ROMANES, ETHEL, *The Life and Letters of George John Romanes* (London: Longmans, Green & Co., 1898)

ROMANES, GEORGE J., *Animal Intelligence* (New York: Appenton & Co., 1884)

——— *A Candid Examination of Theism by Physicus* (Boston: Houghton, Osgood & Co., 1878)

——— *Christian Prayer and General Laws* (London: MacMillan & Co, 1874)

——— *Darwin and After-Darwin: An Exposition of the Darwinian Theory and a Discussion of Post-Darwinian Questions*, 3 vols (Chicago: The Open Court Publishing Company, 1892)

——— 'The Fallacy of Materialism: I. Mind and Body', *Nineteenth Century*, 12 (1882), 871–88

——— *Mental Evolution in Animals: With a Posthumous Essay on Instinct by Charles Darwin* (London: Paul, Trench & Co., 1883)

——— *Mind and Motion and Monism* (London: Longmans, Green & Co, 1895)

——— 'Mr. Butler's *Unconscious Memory*', review, in *Nature*, 23.587 (1881), 286–87

——— 'Mr. Wallace on Darwinism', *Contemporary Review*, 56 (1889), 244–58

——— 'Physiological Selection: An Additional Suggestion on the Origin of Species', *The Journal of the Linnean Society, Zoology*, 19, (1886), 337–411

——— *The Scientific Evidences of Organic Evolution* (London: Macmillan and Co., 1882)

——— *Thoughts on Religion* (London: Longmans, Green & Co., 1904)

RYLANCE, RICK, *Victorian Psychology and British Culture 1850–1880* (Oxford: Oxford University Press, 2000)

[ANON.], 'Samuel Butler', *The Athenaeum*, 3896 (28 June 1902), 819–20.

Samuel Butler: Records and Memorials, ed. by Richard A. Streatfeild (Cambridge: printed for private circulation, 1903)

Samuel Butler, Victorian against the Grain: A Critical Overview, ed. by James Paradis (Toronto: University of Toronto Press, 2007)

SCHACTER, DANIEL L., *Forgotten Ideas, Neglected Pioneers: Richard Semon and the Story of Memory* (Philadelphia: Psychology Press, 2001)

SCHAFFER, SIMON, *From Physics to Anthropology and Back Again* (Cambridge: Prickly Pear Pamphlets, 1994)

SCHWARTZ, JOEL S., *Darwin's Disciple: George John Romanes, a Life in Letters* (Philadelphia: American Philosophical Society, 2010)

——— 'George John Romanes's Defence of Darwinism: The Correspondence of Charles Darwin and his Chief Disciple', *Journal of the History of Biology*, 28.2 (1995), 281–316

——— 'Out from Darwin's Shadow: George John Romanes's Efforts to Popularize Science in "Nineteenth Century" and Other Victorian Periodicals', *Victorian Periodicals Review*, 35.2 (2002), 133–59

Science in the Marketplace: Nineteenth-Century Sites and Experiences, ed. by Aileen Fyfe and Bernard Lightman (Chicago: University of Chicago Press, 2007)

Science Serialized: Representations of the Sciences in Nineteenth-Century Periodicals, ed. by Geoffrey Cantor and Sally Shuttleworth (Cambridge: MIT Press, 2004)

SECORD, JAMES, 'How Scientific Conversation Became Shop Talk', *Science in the Marketplace: Nineteenth-Century Sites and Experiences*, ed. by Aileen Fyfe and Bernard Lightman (Chicago: University of Chicago Press, 2007), pp. 23–59
—— 'Knowledge in Transit', *Isis*, 95 (2004), 654–72
—— *Victorian Sensation: The Extraordinary Publication, Reception, and Secret Authorship of 'Vestiges of the Natural History of Creation'* (Chicago: The University of Chicago Press, 2000)
SHAFFER, ELINOR, *Erewhons of the Eye: Samuel Butler as Painter, Photographer, and Art Critic* (London: Reaktion Books, 1988)
SHAW, GEORGE BERNARD, *Major Barbara* (London: The Court Theatre, 1907)
SHUTTLEWORTH, SALLY, 'Evolutionary Psychology and *The Way of all Flesh*', in *Samuel Butler, Victorian against the Grain: A Critical Overview*, ed. by James Paradis (Toronto: University of Toronto Press, 2007), pp. 21–45
—— *George Eliot and Nineteenth-Century Science: The Make-Believe of a Beginning* (Cambridge: Cambridge University Press, 1984)
SMEATON, W. A., 'Centenary of the Law of Octaves', *Journal of the Royal Institute of Chemistry*, 88 (1964), 271–74
SPENCER, HERBERT, *An Autobiography*, 2 vols (New York: Appleton, 1904)
—— *First Principles of a New System of Philosophy* (New York: Appleton and Co., 1864)
——, THE HAYTHORNE PAPERS, NO. 2: THE DEVELOPMENT HYPOTHESIS', *The Leader*, 3.104, 20 March 1852, pp. 280–81
—— *Principles of Biology* (London: Williams and Norgate, 1864)
—— *Principles of Psychology* (London: Longman, Brown, Green and Longmans, 1855)
—— 'Progress: Its Law and Cause', in Herbert Spencer, *Essays, Scientific, Political, and Speculative* (New York: Appleton, 1892), 8–63
STEBBINS, ROBERT E., 'France', in *The Comparative Reception of Darwinism*, ed. by Thomas E. Glick (Chicago: Chicago University Press, 1988), pp. 117–67
STILLMAN, CLARA, *Samuel Butler: A Mid-Victorian Modern* (New York: The Viking Press, 1932)
STOTT, REBECCA, 'Darwin's Barnacles: Mid-Century Victorian Natural History and the Marine Grotesque', in *Transactions and Encounters: Science and Culture in the Nineteenth Century*, ed. by Roger Luckhurst and Josephine McDonagh (Manchester: Manchester University Press, 2002), pp. 151–82
SULLY, JAMES, 'The Philosophy of Pessimism', *Westminster Review*, 207 (1876), 59–78
SWAIN, EMMA E., 'St. George Mivart as Popularizer of Zoology in Britain and America, 1869–1881', *Endeavour*, 41.4 (2017), 176–91
THOMSON, WILLIAM, *Mathematical and Physical Papers* (Cambridge: Cambridge University Press, 2011)
TOMKEIEFF, S. J., 'The Mnemic Theories of Evolution', *Scientia*, 34 (1923), 159–72
TOPHAM, JONATHAN, 'Publishing "Popular Science" in Early Nineteenth-Century Britain', in *Science in the Marketplace: Nineteenth-Century Sites and Experiences*, ed. by Aileen Fyfe and Bernard Lightman (Chicago: University of Chicago Press, 2007), pp. 135–68
Translators through History, ed. by Jean Delisle and Judith Woodsworth (Philadelphia: John Benjamins Publishing Company, 1995)
TURBIL, CRISTIANO, 'Memory, Heredity and Machines: From Darwinism to Lamarckism in Samuel Butler's *Erewhon*', *Journal of Victorian Culture* (2019) <https://doi.org/10.1093/jvcult/vcz038> [accessed 19.11.2019]
—— 'Making Heredity Matter: Samuel Butler's Idea of Unconscious Memory', *Journal of the History of Biology*, 51, (2018) pp. 7-29.
—— 'In between mental evolution and unconscious memory: Lamarckism, Darwinism and professionalism in late Victorian Britain', *Journal of the History of the Behavioral Sciences*, 53 (2017), 347– 363.

TURNER, R. S., *In the Eye's Mind: Vision and the Helmholtz-Hering Controversy* (Princeton: Princeton University Press, 1994)

—— 'Vision Studies in Germany: Helmholtz versus Hering', *Osiris*, 8 (1993), 80–103

VAN RIPER, A. BOWDOIN, *Men Among the Mammoths* (Chicago: Chicago University Press, 1993)

VAN SPRONSEN, J. W., 'One Hundred Years of the "Law of Octaves": When the Italian Cannizzaro Was Fighting for Atomic Weights in Karlsruhe, Newlands Fought for the Liberation of Italy', *Chymia*, 11 (1966), 125–37

The Victorians and Italy: Literature, Travel, Politics and Art, ed. by Alessandro Vescovi, Luisa Villa and Paul Vita (Monza: Polimetrica, 2007)

VON HARTMANN, EDUARD, *Philosophie des Unbewussten: Versuch einer Weltanschauung* (Berlin: Duncker, 1869)

WALLACE, ALFRED RUSSEL, REVIEW OF SAMUEL BUTLER, *Evolution, Old and New*, in *Nature*, 20, (1879), 141–44

WARD, JAMES, 'Review of Mental Evolution in Animals by G. J. Romanes', *The Athenaeum*, 2773 (1884), 282–83

WARREN, T. H., *A Selection from the Poems of George John Romanes* (London: Longmans, 1986)

WHITE, PAUL, *Thomas Huxley: Making the 'Man of Science'* (Cambridge: Cambridge University Press, 2003)

WILLEY, BASIL, *Darwin and Butler: Two Versions of Evolution. The Hibbert Lectures of 1959* (London: Chatto & Windus, 1960)

ZELLER, PETER, *Romanes: Un discepolo di Darwin alla ricerca delle origini del pensiero* (Rome: Armando Editore, 2007)

ZEMKA, SUE, '*Erewhon* and the End of Utopian Humanism', *ELH*, 69.2 (2002), 339–49

INDEX

'Anti-Victorian' 38–39, 66
Allaman, George J. 29
Allen, Grant 55, 67, 74–75, 105, 110
Angelo, Giovanni 129
Aquinas, Thomas 95
Augustine of Hippo 95

Barlow, Nora 65–66, 83
Baumann, C 32
Barreau, Hervé 31
Barsanti, Giulio 32, 115
Barton, Ruth 114, 116
Beer, Gillian 3–4, 40–45
  'Darwin and Romanticism' 4
  'Prophecy about the future' 42
Bergson, Henri-Louis 15, 25
Bernard, Claude 4, 15, 41, 128
Bigoni, Francesca 115
Bowler, Peter 3, 13
Breuer, Josef 23
Brown, Janet 17
Browning, Robert 129
Bruni, Giuseppe 127
Buffon 5, 38, 72–73, 85, 100, 105
Burrows, Mark S. 115
Butler, Samuel:
  'A Psalm of Montreal' x, xi
  'Darwin among the Machines' 27, 43–49
  'God the Known and God the Unknown' 9, 35, 40, 51–56
  'L'origine Siciliana dell'Odissea' xi
  'Lucubratio Ebria' 23, 46, 47
  'Man of science' 45, 74, 88, 89
  'Remarks on G. J. Romanes' Mental Evolution in Animals' 10, 28, 80, 93, 104–13
  'The Book of the Machines' 27, 35, 43, 49
  'The Germs of Erewhon and of Life and Habit' 48
  A Clergyman's Doubts x
  A First Year in Canterbury Settlement ix
  Alps and Sanctuaries of Piedmont and the Canton Ticino 10, 37, 79, 122, 123
  Authoress of the Odyssey 62, 124
  Erewhon, or, Over the Range 9, 17, 27, 35, 37, 39, 40–54, 64–70, 89, 103, 109, 122, 125, 128
  Evolution, Old and New 10, 17, 20, 28, 55, 62–77, 90, 99–101
  Ex Voto: An Account of the Sacro Monte or New Jerusalem at Varallo-Sesia xi, 123
  Gavottes, Minuets, Fugues: And Other Short Pieces for the Piano xi
  Life and Habit 5, 8, 9, 18, 20, 21, 24, 25, 27–30, 35, 43–54, 70–73, 87–89, 90, 96–97, 102–04, 109–12, 125
  Luck, or Cunning? 6, 17, 18, 22, 27–30
  'On Memory as a Key to the Phenomena of Heredity' 16
  'Remarks on George Romanes' Mental Evolution' 10, 23, 26, 80, 104–13
  Selections from Previous Works 11, 104
  Shakespeare's Sonnets Reconsidered xi
  The Evidence for the Resurrection of Jesus Christ as Contained in the Four Evangelists Critically Examined ix
  The Humour of Homer xi
  The Life and Letters of Dr. Samuel Butler xi
  The Way of all Flesh 18, 37, 39, 42, 122, 128
  Unconscious Memory 8, 9, 10, 15–19, 28, 29–30, 54–55, 76, 79, 89–97, 101, 104–05, 110–13
Butler, Thomas 42

Cantor, Geoffrey 31, 114
Carpenter, William Benjamin 14, 22, 25–27, 50, 100
Carroy, Jacqueline 32
Ceccarelli, David 130
Chambers, Robert 2, 27, 41, 97
Chapman, John 98
Christchurch 4, 49, 67, 81
Cohen, Philip 37, 56
Coleridge, Samuel Taylor 98
Cope, Edward 128
Corsi, Pietro 32

Danziger, Kurt 13, 14
Darwin, Charles:
  'Preface and "A Preliminary Notice"' in Erasmus Darwin by Ernst Krause 17, 64–66, 77–79
  On the Origin of Species by Means of Natural Selection 2, 3, 4, 5, 16, 22, 27, 40
  On the Various Contrivances by which British and Foreign Orchids are Fertilised by Insects ix
  The Descent of Man, and Selection in Relation to Sex x
  The Expression of Emotions in Man and Animals x
  The Variation of Animals and Plants under Domestication ix, 17, 107
Darwin Correspondence Project 70, 71, 81–83
Darwin, Emma 68

Darwin, Erasmus 5, 17, 38
Darwin, Francis 66–72, 80, 126
de Lanessan, Jean-Louis 24
de Quatrefages de Bréau, Jean Louis Armand 24
de Vries, Hugo 107
de Wespin, Jan 123
Delage, Yves 24
Delboeuf, Joseph 25
Delisle, Jean 90
Desmond, Adrian J. 3, 22
Dickens, Charles 87
Dionisi, Antonio 127
Donatello 50

Eliot, George 3, 87
Engels, Eve-Marie 3
Enriques, Federigo 127

Fichte, Johann Gottlieb 23
Figuier, Louis 4
Fitzgerald, F. Scott 37
Forsdyke, Donald R. 18, 125
Fraser, Hilary 129
Furbank, P. N. 39
Fyfe, Aileen 85–86, 114

Giardina, Andrea 127
Gillet, Louis 125
Gillott, David 1, 6, 79
Glick, Thomas F. 3
Goethe, Johann Wolfgang 4, 22–23
Gooch, Joshua A. 45
Grant, Robert 97
Grendon, Felix 52
Gullin, Vincent 25

Haast, John (Julius) 67–69
Hajdenko-Marshall, C. 118
Haeckel, Ernst 22–23, 73, 113
Handel, George Frideric 36, 122
Hardy, Thomas 3
Harkness, Stanley Bates 39, 121
Hartog, Marcus 10, 126–28
Harvey, E. Newton 29
Harvey, Joy 115
Hawkins, Mike 99
Henslow, George 87
Hering, Ewald 5–10, 14– 16, 23–30, 65, 90, 102–04, 106–09, 112–13, 126–28
Hofstadter, Richard 117
Holt, Lee 37, 39
Holterhoff, K. 118
Hooker, Joseph Dalton 67–68
Houghton, William 87
Howard, Daniel 119
Hume, David 15

Huxley, Julian 13
Huxley, Thomas 20, 29–30, 39, 66, 72, 74, 78–79, 85, 87–88, 92, 93, 95

Irvine, William 38–39

Jablonka, Eva 32, 57
Janko, Jan 32
Johns, Charles Alexander 87
Jones, Henry Festing 38, 48, 62, 65, 66, 70, 72, 77, 78, 80
Jones, Steve 10
Jordanova, Ludmilla 32

Kant, Immanuel 15, 23
Kingsley, Charles 3, 87, 112–13
Knowles, James 108
Kohn, David 3
Krause, Ernst 17, 64, 76–80

Lamarck, Jean-Baptiste 3, 5–9, 16, 19–21, 22, 24, 26, 27, 28, 30, 46, 47, 65, 71–76, 85, 92, 95, 97, 98, 100, 101, 104–09, 112, 115
    Histoire naturelle des animaux sans vertèbres 22
    Philosophie Zoologique 21, 46, 108
Lamb, Mariot J. 32, 57
Lanaro, Giorgio 117
Lankester, Ray E. 23, 113
Larbaud, Valery 125
Le Dantec, Félix 24
Leibniz, Gottfried Wilhelm 22
Levine, George 3
Lo Schiavo, Renato 123
Locke, John 15
Lyell, Charles 3, 13, 41
Lytton, Edward Bulwer 42

Mach, Ernst 14
Malthus, Thomas 4
Martins, M. 73
Matthew, Patrick 73, 112
Maudsley, Henry 25, 27
Middleton, Darren J. N. 58
Mill, John Stuart 108
Milton, John 4
Mivart, George 6, 25, 27, 67, 71, 91–98, 99, 101
Moore, James R. 10
Morris, Francis Orpen 87
Morris, William 87
Mucchielli, Laurent 117
Muggeridge, Malcolm 38
Müller, Johannes 23
Myers-Shaffer, Christina 36

Naturphilosophie 22
New Zealand 4, 27, 42, 47–49, 61

Newlands, John 17
Newman, John Henry 58, 115
Nicolas, Serge 32

O'Connor, Ralph 2
Odysseus 123
On the Origin of Species, see Darwin, Charles
Østermark-Johansen, Lene 129
Otis, Laura 13, 18–19, 25
Ovid 4

Paley, William 4, 73, 108–09
Paradis, James 1, 6, 18, 36–40, 61–66, 79
Patmore, Coventry 129
Pauli, Charles Paine 9–11
Pauly, Philip J. 105–06
Perkin, James Russell 58
Poulton, Edward Bagnall 107–09
Protoplasm 13, 29–30

Queen Victoria 2

Rabinbach, Anson 14
Raby, Peter 6, 36, 62
Rachels, James 11
Radick, Gregory 8
Rendall, Vernon Horace 64
Ribot, Théodule 6, 8, 15, 18, 19, 22, 24, 25, 27, 28, 102, 104, 126, 128
Richards, Robert 8
Rignano, Eugenio 10, 126–28
Robinson, Roger 27
Romanes, Ethel 118
Romanes, George 6, 10, 14 (remove dates), 15, 22, 23–26, 104–13
Rossetti, Christina 129
Rylance, Rick 8

Saint-Hilaire, Étienne Geoffroy 8, 29
Savage, Eliza Mary Ann 72, 109
Schacter, Daniel L. 19, 112
Schaffer, Simon 13
Schelling, Friedrich Wilhelm Joseph 23
Schwartz, Joel S. 106–08
Scientia 10, 126–29
Secord, James 2, 40, 41
Sella, Attilio 123, 129
Shaffer, Elinor 6, 26, 39, 122
Shaw, George Bernard 36, 37
Shuttleworth, Sally 3, 18
Smeaton, W. A. 31

Spencer, Herbert 6, 15, 22, 25, 26, 30, 91, 95, 98–104, 110
Stebbins, Robert E. 115
Stillman, Clara 55
Streatfeild, Richard A. 56, 58, 80
Stoker, Bram 87
Stott, Rebecca 11
Suarez, Francisco 94
Sully, James 15, 21
Swain, Emma E. 116
Swift, Jonathan 55

Taine, M. 102
'The European debate' 20–26
The Linnean Society of London 107
The Press 4, 27, 48, 49, 67, 69
Thomson, William 44
Tomkeieff, Sergei J. 128
Topham, Jonathan 86, 114
Turbil, Cristiano 57
Turner, R. S. 32
Tyndall, John 87

Van Riper 11
Van Spronsen, J. W. 31
Verne, Jules 4
Vescovi, Alessandro 129
'Victorian Psychology' 6–9, 13–15, 18–25, 55, 88, 98–101, 110
Villa, Luisa 129
Vita, Paul 129
Von Baer, Karl Ernst Ritter 98
Von Hartmann, Eduard 21, 74
Von Helmholtz, Hermann 23
Von Mohl, Hugo 29

Wallace, Alfred R. 67, 72, 75, 85, 87, 92, 94, 95, 101, 126
Ward, James 15
Warren, T. H. 140
Webb, Thomas Willian 87
Weismann, August 107
Wells, H. G. 87
White, Paul 20, 87, 88
Willey, Basil 64
Woodsworth, Judith 90
Woolf, Virginia 37
Wundt, Wilhelm Maximilian 13, 14

Zeller, Peter 118
Zemka, Sue 43

www.ingramcontent.com/pod-product-compliance
Lightning Source LLC
LaVergne TN
LVHW061252060426
835507LV00017B/2044